Psychoanalyzing the Politics of the New Brain Sciences

Robert Samuels

Psychoanalyzing the Politics of the New Brain Sciences

Robert Samuels
UCSB
Goleta, CA, USA

ISBN 978-3-319-71890-3   ISBN 978-3-319-71891-0   (eBook)
https://doi.org/10.1007/978-3-319-71891-0

Library of Congress Control Number: 2017962904

This Palgrave Pivot imprint is published by Springer Nature
The registered company is Springer International Publishing AG
The registered company address is: Gewerbestrasse 11, 6330 Cham, Switzerland

To Sophia and Madeleine

CONTENTS

CHAPTER 1

Introduction

Abstract This introduction outlines the main arguments of each of the chapters. One of the central points is to show how psychoanalysis offers a critical perspective on the new brain sciences of neuroscience, evolutionary psychology, and behavioral economics. I argue that Freud's basic insights into human subjectivity reveal how instincts are replaced by drives, why humans are not dominated by evolution, why people participate in their own self-destruction, how the mental can disrupt the physical, and why the evolutionary goal of biological survival is often subverted. Although it would be wrong to reject the importance of biology and evolution for human beings, it is equally wrong to believe that we are determined solely by biological forces derived from natural selection. Unfortunately, powerful interests in the world want to convince people that genes and neurotransmitters shape who we are, and so the only solution to many of our psychological and social problems is some form of prescribed medication.

Keywords Psychoanalysis • Neuroscience • Evolutionary psychology • Drives • Freud • Prescription medication

This book argues that neuroscience, evolutionary psychology, and behavioral economics often function as a political ideology masquerading as a new science. In looking at works by Antonio Damasio, Steven Pinker, Richard Thaler, Cas Sunstein, and John Tooby, I offer a close reading of

© The Author(s) 2017   1
R. Samuels, *Psychoanalyzing the Politics of the New Brain Sciences*,
https://doi.org/10.1007/978-3-319-71891-0_1

the new brain sciences, and by turning to the works of Freud and Lacan, I offer a counter-discourse to these new emerging sciences. We shall see that an unintentional political manipulation of scientific thinking serves to repress the psychoanalytic conception of the unconscious and sexuality as it reinforces neoliberalism and promotes the drugging of discontent.

Psychoanalysis helps us to understand how instincts are replaced by drives, why humans are not dominated by evolution, why people participate in their own self-destruction, how the mental can disrupt the physical, and why the evolutionary goal of biological survival is often subverted. Although it would be wrong to reject the importance of biology and evolution for human beings, it is equally wrong to believe that we are determined solely by biological forces derived from natural selection. Unfortunately, powerful interests in the world want to convince people that genes and neurotransmitters shape who we are, and so the only solution to many of our psychological and social problems is some form of prescribed medication.[1]

As I will argue in Chap. 2, neuroscientists like Antonio Damasio return compulsively to the idea that human beings are shaped by inherited instinctual structures centered on promoting self-survival and self-regulation.[2] Even when these scientists allow for environmental influence (nurture) through the theories of neuroplasticity and epigenesis, they often rely on an underlying naturalization of political ideologies. For instance, Damasio's focus on self-regulation and self-survival can be shown to be driven by the political belief in the primacy of self-interested individuals and economic competition.[3] Since genetic evolution depends on mutations, random recombinations, and changing environmental systems, the myth that we are driven by biology and selfish genes to only look out for ourselves and the people who are closest to us can be read as a naturalized political ideology.[4] Here we see how seemingly objective and neutral scientific findings are shaped by particular social and political values and beliefs.

One of the main ways that science is influenced by ideology is through the use of particular words and metaphors. For example, when scientists discuss "cellular machinery" or the idea that the brain is an "information-processing machine," the human mind and body is transformed into a machine or computer.[5] Scientists may say that these are just symbolic terms or convenient placeholders, but as psychoanalysis shows, the choice of words and metaphors does affect how people think and feel about scientific theories.[6] In fact, the very term "selfish gene" pushes people to conceive of genetics as a process privileging the isolated, greedy individual.[7] Although scientists want to hold on to the idea that language is a transparent medium that does not affect the neutrality of scientific research, psychoanalysis helps us to see how unconscious symbolic associations affect scientists and the receivers of scientific knowledge. Therefore, in order to trace the ways particular uses of language affect science and the communication of scientific ideas, I will focus closely on reading the words and metaphors of contemporary brain scientists. One of my goals here is to show the importance of the humanities and psychoanalysis in enhancing our understanding of science and contemporary political ideology.

By paying close attention to the words and arguments of particular theorists, I hope to provide an example of the close reading of scientific texts. This process requires an extensive use of direct quotations so that it is clear that the analysis is derived from the actual words of the source. Moreover, I often examine the rhetoric of the texts from a psychoanalytic and political perspective, which challenges the supposed neutrality of science and other academic disciplines. However, instead of psychoanalyzing the authors, my goal is to provide an analytic account of the social and subjective motivations behind particular shared concepts and theories. Although it may appear that I am attacking these theorists for their destructive intentions, my desire is to explore the ways that ideology shapes unconscious representations. I thus turn to psychoanalysis not only to provide an alternative discourse but to use psychoanalysis as a method of interpretation.

I also want to stress that I do not provide a comprehensive survey of the new brain sciences and psychoanalysis. In contrast to this usual academic process, I focus on close readings of specific texts and theorists, and I point in the notes to other examples of the general argument. In the case of my use of psychoanalysis, I concentrate on employing fundamental concepts and theories derived from Freud and Lacan, and I avoid spending much time in examining the internal debates within the field. At all times, I seek to determine what psychoanalysis can bring specifically to the discussion that other disciplines cannot provide. From my perspective, it is the key concepts of the unconscious, drives, transference, and the superego that establish psychoanalysis as a coherent and separate discourse.

Although much ink has been spilled on the question of whether psychoanalysis is a science or not, I argue that psychoanalysis presents a radical questioning of scientific theory and practice from within science itself. In looking at the language and rhetoric of scientific discourse, I highlight the limits of empirical research, and I question the ability of science to be produced and communicated in a value-free way, and by returning to the subjectivity of the scientists, I hope to unveil the ideology of what I call "neuroliberalism."

What few people have noticed is that the new discourses of neuroscience, evolutionary psychology, and behavioral economics are not simply promoting a particular set of theories, but they are also actively trying to undermine other discourses and disciplines. As I show in Chap. 3, in order to affirm the non-conscious nature of most brain activities, genetic processes, and mental functions, these new sciences often have to debase and attack the social sciences, the humanities, and psychoanalysis.[8] From the perspective of many evolutionary psychologists, they are doing real research based on real facts, while the other academic disciplines are simply making things up and doing unnecessary work. Not only does someone like Steven Pinker feel that most academic disciplines are wrong and silly because they do not affirm biological determinism, he also thinks they are dangerous.[9] What is ironic is that at the same moment that Pinker denies the importance of culture and language in shaping human behavior, he obsesses about how the humanities and the human sciences represent a false cultural view of the world.

Chapter 4 applies this critique of neuroscience and evolutionary psychology to the field of behavioral economics. In looking at the book *Nudge*, I show how misguided lessons learned from neuroscience and evolutionary psychology are now shaping public policies and economic discourse.[10] I also examine how the combination of these new brain sciences represses psychoanalysis and rationalize a social order based on irrational competition, naturalized hierarchy, and political defeatism. I call this new political and social ideology neuroliberalism in order to stress how brain science is being used to justify the neoliberal privatization of public institutions, the promotion of free market fundamentalism, and the erosion of social trust.[11]

As a way of showing what humanists and psychoanalysts can bring to scientific discourse, in Chap. 5, I perform a close reading of several related experiments in evolutionary psychology.[12] In looking at how this research is structured, implemented, and analyzed, I reveal a series of underlying political assumptions shaping this discipline. Ironically, at the very moment evolutionary psychology argues that our mental and social responses to current political issues are inherited through natural selection, we discover that the theory of evolutionary psychology is itself a political social construction that at times attempts to naturalize the dismantling of the welfare state.[13]

The final chapter explains how the rise of neuroliberalism can be understood through the combination of governmental agencies, university research, medical theories, and pharmaceutical solutions in what I call the Governmental University Medical Pharmaceutical Complex (GUMP). Although I do not see this as an intentional conspiracy, the combination of these different social entities results in the notion that all mental problems and social discontent should be dealt through pharmacological solutions. I will argue that the main issue with the GUMP complex is not that people are making huge sums of money selling a false medical paradigm; rather, the problem is that millions of people are having their lives destroyed by ineffective drug regimes.[14] Moreover, a destructive neoliberal ideology is being fueled by these false sciences: since people are told that they are controlled by their genes and neurotransmitters, there appears to be no reason to support progressive parenting, social welfare policies, and education.[15] The paradox is that these neo-conservative ideologies are being produced at supposedly liberal institutions, like public universities.[16] In fact, pharmaceutical companies like to test their new drugs at universities because these institutions are supposed to produce pure research that is objective, neutral, and unbiased. However, the more university scientists have to rely on outside entities to fund their research, the more their research can be corrupted.

One reason why universities have been forced to turn to outside funders to do research is that these institutions have been systematically defunded by state governments.[17] Thus, in order to keep their research process going, public universities have become reliant on the values and priorities of corporations and governmental concerns.[18] Through the grant-funded research process, scientists are now dependent on the outside entities that support their research, and like the corruption of political campaigns by financial contributions, university research is being influenced by special interests.

As I note throughout this book, Freud anticipated many of the problems that we are now dealing with in relation to the medicalization of psychology.[19] He warned against forcing psychoanalysts to be medical doctors because he realized that physicians often are guided by questionable incentives. Furthermore, while Freud always saw himself as a scientist and did hold out a belief that biology would someday resolve the mysteries of the human mind, his own practice and theories offered a radical alternative to the brain sciences. Through his theories of the drive, the unconscious, transference, and the super-ego, Freud presented a coherent challenge to the application of evolutionary biology to the human mind.[20] Moreover, the psychoanalytic practice that Freud developed and Lacan clarified provides an important counter-discourse to neuroscience, evolutionary psychology, and behavioral economics. Unfortunately, the new brain sciences that I will be discussing in the book all tend to dismiss the practice and theory of psychoanalysis.

NOTES

1. Whitaker, Robert. "Anatomy of an epidemic: Psychiatric drugs and the astonishing rise of mental illness in America." *Ethical Human Sciences and Services* 7.1 (2005): 23–35.
2. Damasio, Antonio R. *Descartes' error*. Random House, 2006.
3. Smith, Barbara Herrnstein. "Sewing up the mind: the claims of evolutionary psychology." *Alas, poor Darwin: Arguments against evolutionary psychology* (2000): 129–143.
4. Woolard, Kathryn A., and Bambi B. Schieffelin. "Language ideology." *Annual review of anthropology* (1994): 55–82.
5. Keller, Evelyn Fox. *Refiguring life: Metaphors of twentieth-century biology*. Columbia University Press, 1995.
6. Ogden, Thomas H. "Rediscovering psychoanalysis." *Psychoanalytic Perspectives* 6.1 (2009): 22–31.
7. Dawkins, Richard. *The selfish gene*. Oxford university press, 2016.
8. Rose, Hilary. "Colonising the social sciences." *Alas, poor Darwin: Arguments against evolutionary psychology* (2000): 106–128.
9. Pinker, Steven. *The blank slate: The modern denial of human nature*. Penguin, 2003.
10. Leonard, Thomas C. "Richard H. Thaler, Cass R. Sunstein, Nudge: Improving decisions about health, wealth, and happiness." *Constitutional Political Economy* 19.4 (2008): 356–360.
11. Harvey, David. *A brief history of neoliberalism*. Oxford University Press, USA, 2007.
12. Petersen, Michael Bang, et al. "Who deserves help? evolutionary psychology, social emotions, and public opinion about welfare." *Political psychology* 33.3 (2012): 395–418.
13. Rose, Hilary, and Steven Rose. *Alas, poor Darwin: Arguments against evolutionary psychology*. Random House, 2010: 21.
14. Breggin, Peter Roger, and David Cohen. *Your drug may be your problem: How and why to stop taking psychiatric medications*. Da Capo Press, 2007.

15. Verhaeghe, Paul. *What about Me?: the struggle for identity in a market-based society.* Scribe Publications, 2014.
16. Washburn, Jennifer. *University, Inc.: The corporate corruption of higher education.* Basic Books, 2008.
17. Oliff, Phil, et al. "Recent deep state higher education cuts may harm students and the economy for years to come." *Center on Budget and Policy Priorities* (2013).
18. Kirp, David, and Einstein Shakespeare. "the Bottom Line: The Marketing of Higher Education." (2003): p185.
19. Freud, Sigmund. *The question of lay analysis: Conversations with an impartial person.* WW Norton & Company, 1969.
20. Shepherdson, Charles. *Vital signs: Nature, culture, psychoanalysis.* Psychology Press, 2000.

CHAPTER 2

Damasio's Error: The Politics of Biological Determinism After Freud

Abstract The central argument of this chapter is that neuroscientists like Antonio Damasio rely on an underlying theory of evolution that is full of unrecognized political assumptions and effects. As a prime example of neuroliberalism, Damasio's *Descartes' Error* attempts to naturalize a theory of nature, which is itself in part a product of political ideology and social negotiation. Although he does point to non-biological aspects of the human mind, we shall see that his formulations of consciousness represent a pre-psychoanalytic understanding of subjectivity. Moreover, his focus on survival and adaptation offers a new form of social Darwinism that is determined by a neoliberal emphasis on competitive conformity.

Keywords Damasio • Neuroscience • Social Darwinism • Survival • Liberalism • Adaptation • Neuro-plasticity • Drives

The central argument of this chapter is that neuroscientists like Antonio Damasio rely on an underlying theory of evolution that is full of unrecognized political assumptions and effects. As a prime example of neuroliberalism, Damasio's *Descartes' Error* attempts to naturalize a theory of nature, which is itself in part a product of political ideology and social negotiation.[1] Although he does point to non-biological aspects of the human mind, we shall see that his formulations of consciousness represent a pre-psychoanalytic understanding of subjectivity. Moreover, his focus on survival and adaptation

© The Author(s) 2017
R. Samuels, *Psychoanalyzing the Politics of the New Brain Sciences*,
https://doi.org/10.1007/978-3-319-71891-0_2

offers a new form of social Darwinism that is determined by a neoliberal emphasis on competitive conformity.

The Evolutionary Foundation

At the foundation of Damasio's understanding of the human mind, we find an interpretation of the theory of evolution emphasizing the struggle for individual survival:

> The lower levels in the neural edifice of reason are the same ones that regulate the processing of emotions and feelings, along with the body functions necessary for an organism's survival. In turn, these lower levels maintain direct and mutual relationships with virtually every bodily organ, thus placing the body directly within the chain of operations that generate the highest reaches of reasoning, decision making, and, by extension, social behavior and creativity. Emotion, feeling, and biological regulation all play a role in human reason. The lowly orders of our organism are in the loop of high reason.[2]

Here we see that Damasio wants to show how emotions are tied to reason, and the first way he does this is by insisting on the role of self-regulation and the pursuit of the survival of the individual; however, we have to question the theory of evolution that is being implied in this argument.[3] Even though genetic evolution does point to the survival of certain genetic material, it is unclear if we can equate the survival of particular traits with the survival of the individual. It is also problematic to conflate the self-regulation of a particular organism with the ability of a specific trait to promote survival. In fact, I will argue that the focus on self-regulation and self-survival represents an unacknowledged ideological privileging of individual self-interest in a naturalized competitive environment. Just as neoliberal politics stresses the competitive individual over the cooperative social order, neuroliberalism naturalizes the promotion of the individual through a reductive understanding of survival and self-regulation.

For psychoanalysis, self-regulation is related to the pleasure principle and the idea that the human organism strives to maintain a homeostatic level of reduced tension by avoiding displeasure and pursuing known pleasure.[4] Yet, a key aspect of Freud's work is to show all of the ways that self-regulation breaks down: not only do people pursue self-destructive thoughts and actions, but they seek out levels of enjoyment and pain that disrupt the pleasure principle's efforts to reduce tension to a bare minimum. As Freud insists in his *Beyond the Pleasure Principle*, human beings have a tendency to repeat painful scenes in their dreams and play, and they also tend to repeat the same type of failed relationships, and just when they seem to be getting better in therapy, they seek to leave treatment.[5] Freud tried to combine all of these transgressions of the pleasure principle under the heading of the death drive, and he believed that there was a natural instinct for self-destruction and the return to a primal state of in-animation.[6]

In the case of Damasio's neuroliberalism, a key concept is the notion of adaptation, which he relates to the fundamental function of the human mind:

> Here we have innate circuits whose function is to regulate body function and to ensure the organism's survival, achieved by controlling the internal biochemical operations of the endocrine system, immune system, and viscera, and drives and instincts. Why should these circuits interfere with the shaping of the more modern and plastic ones concerned with representing our acquired experiences? The answer to this important question lies in the fact that both the records of experiences and the responses to them, if they are to be adaptive, must be evaluated and shaped by a fundamental set of preferences of the organism that consider survival paramount. It appears that because this evaluation and shaping are vital for the continuation of the organism, genes also specify that the innate circuits must exert a powerful influence on virtually the entire set of circuits that can be modified by experience.[7]

Implied in this conception of evolutionary adaptation is a utilitarian conception of human beings: the basic idea is that individuals pursue their self-interest in a rational and functional way. Yet, adaptation in evolution only becomes recognizable after the fact when a random mutation has resulted in a transformation that ends up increasing the likelihood of a particular trait surviving into the future. As Richard Dawkins insists, there is no plan or purpose in evolution, and so the use of language implying intention or self-interest is not appropriate.[8]

Damasio's discussion of self-regulation and adaptation for survival then feeds into a political psychology that promotes the rational self-interest of the individual who knows how to adapt to a particular social environment; in other terms, the model for subjectivity promoted by neoliberalism can be called competitive conformity: society rewards people who compete in a capitalist market by following cultural norms and expectations without questioning the value or the ethics of the social order.[9] In fact, when Damasio discusses a famous case of a man who has lost the ability to make ethical judgments, this neuroscientist unintentionally exposes what he values in people:

> Gage had once known all he needed to know about making choices conducive to his betterment. He had a sense of personal and social responsibility, reflected in the way he had secured advancement in his job, cared for the quality of his work, and attracted the admiration of employers and colleagues. He was well adapted in terms of social convention and appears to have been ethical in his dealings. After the accident, he no longer showed respect for social convention; ethics in the broad sense of the term were violated; the decisions he made did not take into account his best interest, and he was given to invent tales "without any foundation except in his fancy," in Harlow's words. There was no evidence of concern about his future, no sign of forethought.[10]

In explaining the negative effects of Gage's brain injury, Damasio reveals that his "neutral" science is biased toward a certain vision of normality: the normal man cares about the quality of his work, is well adapted to social conventions, seeks out the admiration of others, and follows his self-interests. We can consider these social traits to be the product of a particular political vision best suited for upper-middle-class workers who do not seek to challenge the social system. Not only does Damasio propose this defining description without qualification, but his focus on conformity dovetails with his conception of evolutionary adaptation; in fact, I would argue that Damasio naturalizes the conformity of the ideal capitalist neoliberal worker, who is able to pursue his self-interest by adapting to any new situation without causing conflict or rebelling.

This perfect neuroliberal worker is also given a chemical foundation:

> Modulator neurons distribute neurotransmitters (such as dopamine, norepinephrine, serotonin and acetylcholine) to widespread regions of the cerebral cortex and subcortical nuclei. This clever arrangement can be described as follows: (1) the innate regulatory circuits are involved in the business of organism survival and because of that they are privy to what is happening in the more modern sectors of the brain; (2) the goodness and badness of situations is regularly signaled to them; and (3) they express their inherent reaction to

goodness and badness by influencing how the rest of the brain is shaped, so that it can assist survival in the most efficacious way.[11]

The "business" of survival is therefore determined by neurotransmitters that make sure the human machine runs smoothly and efficiently: like a well-drugged worker, the brain is imagined to be powered by chemicals that increase efficiency and avoid bad situations. Damasio does not ask here what happens if the worker-brain wants to be bad or if he does not like his job or he wants to join with others to protest his working conditions.

From a psychoanalytic perspective, the self is made up of multiple, competing agencies or thought processes, and so it is hard to talk about an individual simply adapting to the environment or automatically determining the goodness or badness of a particular situation. On one level, instinctual impulses are transformed into drives and compulsive impulses that are often at odds with cultural norms and individual goals. For instance, someone who is addicted to pornography may feel compelled to consume more representations of violent sexuality, and at the same time, this person may feel guilty and ashamed because society says that it is wrong to objectify and degrade women. In fact, for Freud, one of the main roles of the super-ego is to judge the ego in relation to the cultural ideals of the ego ideal. Here we find three parts of the self at war with each other, and if we add the impulses of the drives to the mix, we see that it is hard to talk about a unified individual seeking to conform to the environment.

Although neuroscientists like Damasio do at times account for competing mental processes, they tend to posit a simple opposition between inherited unconscious mental programs and the conscious rationalization of these adapted formations. Moreover, while psychoanalysis sees repression as the key to the unconscious, neuroscientists usually dismiss this concept and do not account for the ways that people lie to themselves or misrepresent their own perceptions, impulses, and actions to themselves in order to avoid discomfort or conflict. Often neuroscientists can do away with the psychoanalytic notions of the unconscious and repression because they confuse the unconscious with any process that it is simply nonconscious. For example, the following passage presents the human brain and mind as circuits driven by possibly "mindless" inherited instincts and chemicals:

In general, drives and instincts operate either by generating a particular behavior directly or by inducing physiological states that lead individuals to behave in a particular way, mindlessly or not. Virtually all the behaviors ensuing from drives and instincts contribute to survival either directly, by performing a life-saving action, or indirectly, by propitiating conditions advantageous to survival or reducing the influence of potentially harmful conditions. Emotions and feelings, which are central to the view of rationality I am proposing, are a powerful manifestation of drives and instincts, part and parcel of their workings.[12]

One of the curious aspects of this passage is the repetition of the terms "drives and instincts": for psychoanalysts like Lacan, the distinction between animal instincts and human drives is crucial. Lacan argued that the translators of Freud performed a disservice by translating two very different German words, *Trieb* and *Instinkt*, by the same term "instinct."[13] According to Lacan, this is a huge problem because psychoanalysis is in part grounded on the notion that animals have instincts, but humans have drives, and drives represent a break with evolutionary biological determinism. For instance, humans can become addicted to almost anything because unlike animals, they are not dominated by preformed instincts that link an internal need to a specific object in the environment. Thus, one of the things that make us human is that almost anything can be sexualized, including pain and self-destruction.[14] This openness of human drives renders problematic any focus on self-regulation, biological adaptation, or the sole pursuit of survival: in opposition to Damasio, psychoanalysis tells us that the self-interested pursuit of individual survival through the adaptation to a particular environment is constantly being disrupted by sexuality and drives.[15]

Unfortunately, in this text, one of Damasio's only references to Freud reduces psychological conflict to the reductive opposition of instincts versus social regulation:

The creation of a superego which would accommodate instincts to social dictates was Freud's formulation, in *Civilization and Its Discontents*, which was stripped of Cartesian dualism but was nowhere explicit in neural terms. A task that faces neuroscientists today is to consider the neurobiology supporting adaptive supraregulations, by which I mean the study and understanding of the brain structures required to know about those regulations.[16]

From Damasio's neuroscientific perspective, psychoanalysis needs to base its findings in evolution and brain structure in order to understand the process of adaptation and social regulation of instincts. However, for Freud, just as drives undermine biological determinism, the super-ego also challenges the evolutionary theory of successful adaptation. From a Freudian perspective, the super-ego draws its energy from the drives and its content from culture in a structure where the more one conforms to social norms, the stronger and harsher one's conscience becomes.[17] For instance, in the case of obsessional neurosis, Freud found that when a person renounces sexual impulses to adapt to social dictates, the conscience itself becomes sexualized and a source of unconscious enjoyment.[18] One thus feels good about criticizing oneself, and the function of social regulation and adaptation becomes non-adaptive for both the individual and society.

In contrast to Freud's complex understanding of the super-ego and the social regulation of anti-social impulses, Damasio, like so many other neuroscientists, evolutionary psychologists, and behavioral economists, believes that our instincts are shaped by inherited mental programs and our ability to rationally decide to conform to social expectations and norms:

> In human societies there are social conventions and ethical rules over and above those that biology already provides. Those additional layers of control shape instinctual behavior so that it can be adapted flexibly to a complex and rapidly changing environment and ensure survival for the individual and for others ... I see a "trail" connecting the brain that represents one, to the brain that represents the other. Naturally, that trail is made up of connections among neurons.[19]

Here we find a great example of neuroliberalism: cultural variation only functions to allow the adapting subject the ability to adjust to different situations. At first, it appears that Damasio is moving beyond pure biological determinism, but he returns at the end of this passage to the notion that the social can be translated into a system of neurons located in individual brains.

What is missing from this theory is any sense that there are social constructs that cannot be traced back to the workings of individual brains. We have to ask, for example, how does a brain imagining machine measure social interactions that go beyond the minds of isolated individuals? Can a brain scan detect justice or democracy or infinity? Do brain scientists end up privileging the isolated individual because that is the only thing they can really try to measure with a machine? In Damasio's case, any complex social or human concept is shown to be derived from the underlying biological program materialized in neural networks and powered by neurotransmitters:

> For most ethical rules and social conventions, regardless of how elevated their goal, I believe one can envision a meaningful link to simpler goals and to drives and instincts. Why should this be so? Because the consequences of achieving or not achieving a rarefied social goal contribute (or are perceived as contributing), albeit indirectly, to survival and to the quality of that survival.[20]

Since from Damasio's perspective, every human thought and action is driven by the pursuit for survival, individuals are defined by their ability to achieve rarefied social goals. From the view of neoliberalism and neuroscience, we are driven to outcompete others in order to survive as competition takes over all aspects of human experience.

The Neuroliberal Culture of Narcissism

As Christopher Lasch argues in *The Culture of Narcissism*, the contemporary stress on individual survival results in a society where people are socialized to become other-directed; in other terms, the imaginary structure of narcissism becomes the dominant mode of subjectivity, and this means that the isolated individual becomes obsessed by having his or her own ego recognized by a cultural ideal or social Other.[21] Thus, unlike the modern celebration of the rugged individualist, the neuroliberal subject is encouraged to sell its self as a commodity on the market in order to have the ideal ego verified by the ego ideal.[22] In referring to David Riesman, Lasch points out that in an age of rapid job turnover, the narcissistic individual is motivated to adapt to any new situation.[23] Lasch adds that while it looks like this new form of individualism is based on the personal pursuit of pleasure, the truth is that people are being socialized to exploit "the conventions of interpersonal relations for their own benefit."[24] Neoliberalism therefore leads to cynical competitive conformity: one learns how to conform to a broken system and use others to promote one's own survival.

In popular culture, we see many examples of this neoliberal subjectivity. For instance in shows like *Breaking Bad*, *Weeds*, and *The Sopranos*, people engage in criminal activities in order to help their families survive difficult economic times. The cynicism of these shows is evident in the fact that it is clear that the criminals always have a choice to do things differently, but they seem to enjoy their illegal behavior, and they use the survival of the family as an excuse to enhance their own social power. For Lasch, economic inequality and poverty undermine our belief in social institutions, and so we are left with nothing else to do than to focus on protecting our self-survival and the survival of our kin in the present.[25] Making matters worse is that the focus on death and destruction in the media feeds a sense of doom and insecurity, which increases the desire to concentrate on pure survival.

In the case of neuroliberalism, the older social Darwinism centered on the "survival of the fittest" is given a new life in a neoliberal context.[26] Once again, the social hierarchy is naturalized, but instead of focusing on the genetic determination of race and gender, the new social Darwinists offer a more open system in the sense that the fittest people are now defined by how well they adapt selfishly to a changing environment. For example, through the concepts of neuroplasticity and epigenesis, biological determinism is combined with social conformity: according to this neuro-logic, we are preprogrammed to compete and adapt to every new social situation.[27] However, psychoanalysis tells us that since the narcissist is defined by others and what others think of the self, the narcissist has no core inner self, and as Lacan insists, the narcissist acts as an other in order to be verified by an other.[28] Therefore, the neuroscience discourse of survival, self-regulation, and adaptation leads to a culture of neoliberal narcissism: people conform to an unequal competitive environment in order to pursue their perceived individual advantage, but they become alienated from their own selves as they seek to be verified by idealized others.

This focus on a survivalist mentality can also be seen in Naomi Klein's *The Shock Doctrine*, where she shows how the conservative counter-revolution against the welfare state is reliant on the production of constant crisis.[29] Thus, in order to impose austerity measures that privatize public institutions and "liberalize" economies, economists and politicians take advantages of financial and natural crisis as they argue that the only way for a nation to survive in the face of danger is to adapt to the free market and stop using government to regulate businesses and tax wealthy people. Once again, neuroliberalism gives this political strategy a scientific foundation by showing how we are designed by evolution to be selfish, adaptive, and competitive as we conform to a changing environment.

Not only does this turn to science naturalize political ideology, but it also argues that any resistance is futile because we are defined by our universal human nature. Moreover, the use of science to legitimize politics and the status quo allows people to argue that their ideology is not ideological since it is based on empirical facts. Yet, at the foundation of neuroliberalism, we find the socially constructed notion that everything is a market, and markets are a natural way of deciding who wins and who loses in society. At time Damasio does seem aware of the potential risks of neuroliberalism:

> Does this mean that love, generosity, kindness, compassion, honesty, and other commendable human characteristics are nothing but the result of conscious but selfish, survival-oriented neurobiological regulation? Does this deny the possibility of altruism and negate free will? Does this mean that there is no true love, no sincere friendship, no genuine compassion? That is definitely not the case. Love is true, friendship sincere, and compassion genuine, if I do not lie about how I feel, if I really feel loving, friendly, and compassionate. Perhaps I would be more eligible for praise if I arrived at such sentiments by means of pure intellectual effort and willpower, but what if I have not, what if my current nature helps me get there faster, and be nice and honest without even trying? The truth of the feeling (which concerns how what I do and say matches what I have in mind), the magnitude of the feeling, and the beauty of the feeling, are not endangered by realizing that survival, brain, and proper education have a lot to do with the reasons why we experience such feelings.[30]

As Freud or Shakespeare might argue, it does appear that Damasio protests too much here; as he tries to show that he is aware of the limits of biological determinism, he cannot help but argue that even if we are completely determined by nature, that would not mean that our free will or altruism would be less valuable. Furthermore, he argues that honesty is not necessarily based on authenticity or realness; rather, truth is defined as the perfect match between what people say and what they have in mind. Thus, instead of seeing truth as having a social basis or even an empirical foundation, truth has to do with the match between words and intentions.

Of course, psychoanalysis has problematized the correspondence between inner intentions and speech; after all, the unconscious represents an unintentional discourse that surprises the individual.[31] Moreover, what Damasio never considers is the psychoanalytic notion that people lie to themselves and how this form of self-deception opens up the door for repression and unconscious formations. In fact, by excluding repression, neuroscientists, evolutionary psychologists, and behavioral economists are able to posit a cognitive unconscious or non-conscious state that is divorced from Freudian notions of repression, sexuality, and self-defeating behavior.

From a psychoanalytic perspective, the unconscious represents a set of interrelated processes that are not based on intentional control or biological adaptation. According to Freud and Lacan, the unconscious is centered on the fact that even if we try to deny our unpleasant or immoral experiences, a reminder of those events returns in the form of dreams, symptoms, and fantasies. These unconscious formations always represent substitutions and displacements of the original referent, and here we see how the unconscious communicates in an indirect fashion using a form of poetic symbolism. Moreover, unconscious representations center around memories of the interpretation of a particular event from an affective, ethical, and cognitive perspective. For example, if I dream that I had a fight with my dead uncle, we first must understand that my uncle represents someone else and the fight points to a feeling that was derived from an event that I interpreted from a particular social and cultural perspective. As Lacan argues, the domain of the unconscious is ethical because it is centered on my interpretation of how an event or feeling relates to my particular social norms and conventions. However, this affective interpretation occurred in the past and has now been forgotten, but the feelings of the event return in a distorted form in my unconscious dream.

The reason why I am focusing on Freud and Lacan's conceptions of the unconscious, the drives, the super-ego, and narcissism is that I have found that other psychoanalytic theorists do not fully grasp the psychoanalytic understanding of these key concepts. Furthermore, most psychoanalysts only focus on part of the Freudian system and often fall into the trap of not differentiating between the Freudian understanding of these terms and the more common psychological interpretation of these processes. In other terms, the problem is not simply that neuroscientists ignore psychoanalysis, but psychoanalysts also do not comprehend the radical meaning of Freud's theories. One of the reasons for this problem is that Freud often used common terms to point to very uncommon notions. For example, the concept of the unconscious is easily equated with any mental process that is not conscious.

Instead of recognizing the roles of the unconscious, drives, self-destruction, and human sexuality from the start to the end of human development, Damasio refinds the instinct for survival and self-regulation:

> The picture I am drawing for humans is that of an organism that comes to life designed with automatic survival mechanisms, and to which education and acculturation add a set of socially permissible and desirable decision-making strategies that, in turn, enhance survival, remarkably improve the quality of that survival, and serve as the basis for constructing a person.[32]

This focus on survival on all levels of human existence crowds out non-utilitarian and irrational aspects of subjectivity as it also limits the importance of art, play, culture, and self-reflection. As a product of neuroliberalism, the focus is on how humans are biologically determined to only care about their own survival, and even when social goals intervene, they are themselves directed by the same biological mechanisms.

To be fair to Damasio, he does at times indicate that there are levels of human experience that are not directly defined by biological adaptation and survival strategies:

> From an evolutionary perspective, the oldest decision-making device pertains to basic biological regulation; the next, to the personal and social realm; and the most recent, to a collection of abstract-symbolic operations under which we can find artistic and scientific reasoning, utilitarian-engineering reasoning, and the developments of language and mathematics. But although ages of evolution and dedicated neural systems may confer some independence to each of these reasoning/decision-making "modules," I suspect they are all interdependent.[33]

Although he argues that there may be some autonomy for cultural influences, he still sees the biological base as dominating and influencing all aspects of human subjectivity. In fact, it appears that the exceptions to evolutionary control help to conform the rule of biological determinism.

While few people would deny the strong hereditary role in physical traits, the evolutionary determination of mental characteristics and functions is much more debatable. In the following passage, we see how evolutionary psychologists and neuroscientists often join hands in retroactively reconstructing an imaginary history to explain the biological determination of the human mind:

> when brains complex enough to generate not just motor responses (actions) but also mental responses (images in the mind) were selected in evolution, it was probably because those mental responses enhanced organism survival by one or all of the following means: a greater appreciation of external circumstances (for instance, perceiving more details about an object, locating it more accurately in space, and so on); a refinement of motor responses (hitting a target with greater precision); and a prediction of future consequences by way of imagining scenarios and planning actions conducive to achieving the best imagined scenarios.[34]

This theory of how evolution has shaped our mental abilities is forced to turn to an imaginary depiction of the past in order to explain how images help to define human mental functioning.[35]

MISREADING DESCARTES

Damasio's turn to the connection between the body and the mind represents an idealistic scientific attempt to heal what he sees as the Cartesian split between mental reason and bodily emotion:

> It would not have been possible to present my side of this conversation without invoking Descartes as an emblem for a collection of ideas on body, brain, and mind that in one way or another remain influential in Western sciences and humanities. My concern, as you have seen, is for both the dualist notion with which Descartes split the mind from brain and body (in its extreme version, it holds less sway) and for the modern variants of this notion: the idea, for instance, that mind and brain are related, but only in the sense that the mind is the software program run in a piece of computer hardware called brain; or that brain and body are related, but only in the sense that the former cannot survive without the life support of the latter.[36]

Before we problematize this interpretation of Descartes, it is important to stress that Damasio's central project in *Descartes' Error* is to use the unity of the body and the brain as a way of showing how effective reason always relies on emotion. What is considered pathological for Damasio is when one splits off reason from emotion as can be seen in his discussion of patients with particular brain injuries; however, what helps to tie Descartes to psychoanalysis is precisely the effort to separate immaterial reason from the empirical body.

At the start of Descartes' *Discourse on the Method*, he makes the following claim: "Good sense is, of all things among men, the most equally distributed."[37] Of course, one could easily respond to this argument that it is not true and does not match reality, but this democratic definition of reason must be considered to be an impossible ideal that does not exist in nature or the real but is an artificial human construct that makes possible a whole set of institutions, practices, and beliefs. Modern reason, then, begins with a leap of faith, and it is hard to see how this type of social construct can be derived from evolution or registered in a brain scan.

For Descartes and psychoanalysis, the universality of reason, which allows for science and democratic institutions, reveals how humans can go against nature and promote ways of acting and perceiving the world that do not rely on biology, evolution, nature, or self-survival. There is thus an irrational foundation to rationality, and yet without the ideal conception of reason, we cannot even attempt to achieve the scientific and democratic goals of objectivity, neutrality, and universality.[38] Unfortunately, since neuroscientists, evolutionary psychologists, and behavioral economists tend to focus on what can be detected in an isolated individual's brain, they cannot access the ideal social abstractions of reason, and so they unintentionally promote a neoliberal world view that privileges the adaptive individual over abstract social universals, which are necessary for social solidarity and democratic reason.

If we look at what Descartes actually wrote about science and reason, we see how his establishment of the scientific method relies on effacing self-interest and inherited beliefs: "The first [step] was never to accept anything for true which I did not clearly know to be such; that is to say, carefully to avoid precipitancy and prejudice, and to comprise nothing more in my judgment than what was presented to my mind so clearly and distinctly as to exclude all ground of doubt."[39] Descartes' methodological doubt is not only a clear effort at breaking with the premodern stress on faith, fate, and belief, but it also shows how the scientist has to seek out the certainty of material reality through a process of self-erasure. We can call this self-cancellation objectivity and neutrality, but what is so important for Lacan is the idea that the subject is defined as being empty and yet active; here, we see one of the effects of the body being identified with the unified external image as an empty container with a clear inside and outside.[40] The modern subject is then free because it is empty of all content, and it is not determined by biological evolution, cultural tradition, or internalized prejudices.

Once Descartes establishes the impossible but necessary ideal of the empty, universal subject of science, he adds that one of the crucial steps of scientific thinking is the "assigning in thought a certain order even to those objects which in their own nature do not stand in a relation of antecedence and sequence."[41] In other terms, the scientific search for truth requires the imposition of an artificial and unnatural order in the form of logic and math. Since math is a human construct that does not exist in nature, it constitutes a radical break with evolution and biological determinism: the desire to discover the truth of nature ends up relying on an unnatural set of shared fictions, which allow for the human mind to imagine that it can represent nature without prejudice or distortion.

After Descartes establishes the foundation of the modern scientific method, he realizes that if he actually doubted all previous ideas, he would be in a state of total anxiety and insanity, and so he decided to temporarily conform to the people around him. However, in his effort to determine what aspects of other people he should copy, he makes the following declaration:

> I was convinced that I could not do better than follow in the meantime the opinions of the most judicious; and although there are some perhaps among the Persians and Chinese as judicious as among ourselves, expediency seemed to dictate that I should regulate my practice conformably to the opinions of those with whom I should have to live; and it appeared to me that, in order to ascertain the real opinions of such, I ought rather to take cognizance of what they practised than of what they said, not only because, in the corruption of our manners, there are few disposed to speak exactly as they believe, but also because very many are not aware of what it is that they really believe; for, as the act of mind by which a thing is believed is different from that by which we know that we believe it...[42]

Like a good social scientist, Descartes realizes that people may not accurately self-report on their own behaviors, and more problematically, people may not be aware of their own beliefs. In other terms, Descartes realizes that belief itself may be unconscious and not accessible to self-knowledge.

Descartes' vision of the social subject contrasts with his theories of the subject of science and the subject of democracy; while the subject of democracy is a universal subject affirmed through an irrational leap and the subject of science is an empty subject hooked up to an artificial logic, the social subject misrepresents his own actions and beliefs to himself. It should be clear that all three of these different levels of subjectivity challenge Damasio's neuroscientific logic; as Descartes reveals, we are all internally divided and hold multiple and conflicting levels of mental activity. In fact, Descartes adds an additional dimension to his theory of social subjectivity by arguing that social conformity is driven by the desire to escape from the internal awareness of guilt, shame, and remorse: "This principle was sufficient thenceforward to rid me of all those repentings and pangs of remorse that usually disturb the consciences of such feeble and uncertain minds as, destitute of any clear and determinate principle of choice, allow themselves one day to adopt a course of action as the best, which they abandon the next, as the opposite."[43] Descartes argues here that he decided to commit to his social conformity even if it proved to be wrong or excessive because once one makes a choice, one should stick to it no matter what since it delivers one from one's guilty conscience. From this perspective, the key to cynical conformity is that copying the actions of others allows us to escape responsibility and frees us from doubt and guilt. Therefore, even if we may doubt the correctness of the behavior of others or the entire social system, we should conform to escape our own unconscious. Here we find the essence of neoliberal subjectivity: people compete in a system in which they do not believe so that they do not have to feel any guilt or responsibility.

In terms of psychoanalysis, Freud used the concepts of transference, idealization, identification, and narcissism to explain how and why people conform to cultural ideals in order to escape from their own conscience. For instance, when a patient idealizes the analyst as the one who knows and understands the patient's problems and the meaning of the patient's unconscious formations, the patient is idealizing the analyst and identifying with the analyst's knowledge. Here, the analyst becomes the idealized savior who verifies the subject's own sense of knowledge and control, and at the same time, the responsibility for the patient's guilt and shame is handed over to the analyst. In fact, when Lacan explains the psychoanalytic notion of transference, he turns to Descartes' use of god as the ultimate guarantee of truth and knowledge.[44] For Lacan, this idealization of the Other as the one who knows defines transference and the initial stages of analysis where the patient seeks to place the analyst in the ideal position of the ego ideal that verifies and supports the patient's ideal ego. In this formation of a narcissistic relationship, the patient acts like Descartes when he decided to conform to the actions of others in order to escape his own conscience. Transference therefore can help us to understand how and why people conform to social norms in neoliberal culture: by placing others in the position of the ideal Other, one is able to escape one's own guilt, shame, and responsibility, but once the Other fails to acknowledge the subject's ideal ego, resentment emerges and one feels attacked by the harsh, critical super-ego. The analyst then has to maintain analytic neutrality by not playing the role of the ideal Other or the super-ego and refusing to comply with the demands for love and understanding that the patient makes to the analyst in the transference.

WHAT DESCARTES REALLY WROTE

I have been arguing here that Descartes' self-analysis in the *Discourse on Method* can be read as a pre-psychoanalytic discussion of how human beings break with nature and are subject to forms of thinking and social relating that are not purely empirical or material. From this perspective, at the start of modern science, we encounter the limits of science and the need to develop a psychoanalytic understanding of subjectivity and intersubjectivity. The paradox then of much of the neuroliberal brain sciences is that they turn to a reductive understanding of human thinking and behavior as they position a techno-science as the idealized Other that understands our unconscious. For instance, while Damasio fails to recognize Descartes' most important insights into science, human subjectivity, or democratic reason, he returns to a set of reductive, received ideas:

The statement, perhaps the most famous in the history of philosophy, appears first in the fourth section of the Discourse on the Method (1637), in French ("Je pense donc je suis"); and then in the first part of the Principles of Philosophy (1644), in Latin ("Cogito ergo sum"). Taken literally, the statement illustrates precisely the opposite of what I believe to be true about the origins of mind and about the relation between mind and body. It suggests that thinking, and awareness of thinking, are the real substrates of being. And since we know that Descartes imagined thinking as an activity quite separate from the body, it does celebrate the separation of mind, the "thinking thing" (res cogitans), from the nonthinking body, that which has extension and mechanical parts (res extensa).[45]

What Damasio does not indicate in this interpretation of Descartes is that right before Descartes affirms his "I think, therefore, I am," in the *Discourse on Method*, he indicates that there is no way of separating the fictional world of the unconscious dream state from the reality-based world experienced in wakeful consciousness:

> finally, when I considered that the very same thoughts (presentations) which we experience when awake may also be experienced when we are asleep, while there is at that time not one of them true, I supposed that all the objects (presentations) that had ever entered into my mind when awake, had in them no more truth than the illusions of my dreams. But immediately upon this I observed that, whilst I thus wished to think that all was false, it was absolutely necessary that I, who thus thought, should be somewhat; and as I observed that this truth, I think, therefore I am (*cogito ergo sum*), was so certain and of such evidence that no ground of doubt, however extravagant, could be alleged by the sceptics capable of shaking it.[46]

As Lacan stresses, psychoanalysis only becomes possible after Descartes, and one reason is that the fictional world of the unconscious is given the same status as the conscious thinking subject.[47] Since at any single moment, one does not know if one is awake or still dreaming, mental autonomy transcends material reality, and here we find a radical break with nature, evolution, and biological determinism.

THE STORY OF REPETITION

Damasio's error can be seen in how he has to repress Descartes' use of the unconscious, cynical conformity, democratic reason, and scientific neutrality in order to constitute the ideal workings of human brain preprogrammed by evolution to adapt to any particular social situation. Instead of affirming the multiple levels of human subjectivity that we located in Descartes, Freud, and Lacan, neuroscience has to equate any disruption of the human instinct for adaptive survival to a physical illness or injury. This is why neuroscientists tend to repeat the same extraordinary stories of brain-injured patients like Phineas Gage, who was able to reason but not make ethical decisions after an iron rod blasted a hole through his brain.[48]

In place of looking at how people respond to mental trauma or bad objects and feelings by repressing memories into the unconscious, Damasio has to rely on narratives depicting the destruction of particular brain regions to show how specific parts of the brain control particular mental functions. Although Damasio himself often challenges the equation of a specific brain region with a specific mental function, he cannot stop relying on this questionable anatomical argument:

> There are "systems" made up of several interconnected brain units; ana-tomically, but not functionally, those brain units are none other than the old "centers" of phrenologically inspired theory; and these systems are indeed dedicated to relatively separable operations that constitute the basis of men-tal functions ... What determines the contribution of a given brain unit to the operation of the system to which it belongs is not just the structure of the unit but also its place in the system. The whereabouts of a unit is of para-mount importance. This is why throughout this book I will talk so much about neuroanatomy, or brain anatomy.[49]

Damasio is making a difficult and contradictory argument here: he has to affirm that the brain is an integrated system as he still equates particular brain regions with particular mental functions and specific neurotransmit-ters and ultimately specific genetic material. Once again, we could argue that Damasio protests too much, and his efforts to deny that neuroscience represents a new form of phrenology reveals that on a deeper, unconscious level, he is afraid that he is repeating the same mistakes of past scientific efforts to localize mental functions in specific brain regions.

On a fundamental level, neuroscience, evolutionary psychology, and behavioral economics can only work if they first define a particular mental function and then tie it to a particular brain region or set of regions, which are, in turn, associated with specific neurotransmitters related to a specific story of why a particular function was selected by evolution to ensure suc-cessful adaptation. The first step and problem of this process is the defini-tion of the mental function. For instance, here is how Damasio defines reason: "It is perhaps accurate to say that the purpose of reasoning is decid-ing and that the essence of deciding is selecting a response option, that is, choosing a nonverbal action, a word, a sentence, or some combination thereof, among the many possible at the moment, in connection with a given situation. Reasoning and deciding are so interwoven that they are often used interchangeably."[50] This equation of reason with decision-making is a highly reductive understanding of the possible meanings of this mental process. As I argued above, when Descartes defines reason as an equally shared good sense, he is clearly not talking about simply making decisions; rather, reason for Descartes means several things, including the social belief in democratic equality. Yet, for Damasio, reason can be equated with decision-making because he is relying on the idea that normal, rational people pursue their self-interest in order to adapt to a particular environment and enhance their own personal survival. The very definition of a mental function is thus a product of political ideology: the scientist takes it for granted that his definitions of terms are neutral and objective, and yet they are full of bias and assumptions.

Once a mental function is determined, it can be fed into the predetermined biological machine system, which is itself defined by a set of presuppositions and political projections. For example, in the following passage, the objective and neutral discourse belies a hidden agenda:

> (1) The human brain and the rest of the body constitute an indissociable organism, integrated by means of mutually interactive biochemical and neural regulatory circuits (including endocrine, immune, and autonomic neural components); (2) The organism interacts with the environment as an ensemble: the interaction is neither of the body alone nor of the brain alone; (3) The physiological operations that we call mind are derived from the structural and functional ensemble rather than from the brain alone: mental phenomena can be fully understood only in the context of an organism's interacting in an environment. That the environment is, in part, a product of the organism's activity itself, merely underscores the complexity of interactions we must take into account.[51]

This ideal unification of the brain, the mind, and the body appears to make perfect sense, and yet we have to ask if the proposed system really exists in reality or if it is simply a way of retroactively explaining how genes, neurotransmitters, brain regions, and mental functions hook up in a seamless order. At no point in his work does Damasio probe the possible disruptive relation between the brain and the mind or the fact that the mind can be in conflict with the biologically determined brain. Also, he is never able to directly tie particular genes to particular mental functions because it is doubtful that anyone has been able to establish this clear connection.[52] The whole foundation of the evolutionary theory of mental functioning turns out to be based on a series of unacknowledged leaps and associations tied together by convincing rhetoric and storytelling.[53]

SCIENCE FICTIONS

In one very illuminating passage, the narrative nature of this science comes into focus:

> I wrote this book as my side of a conversation with a curious, intelligent, and wise imaginary friend, who knew little about neuro-science but much about life. We made a deal: the conversation was to have mutual benefits. My friend was to learn about the brain and about those mysterious things mental, and I was to gain insights as I struggled to explain my idea of what body, brain, and mind are about. We agreed not to turn the conversation into a boring lecture, not to disagree violently, and not to try to cover too much. I would talk about established facts, about facts in doubt, and about hypotheses, even when I could come up with nothing but hunches to support them.[54]

What is so interested about this passage is the question of how does it frames Damasio's entire discourse. From a psychoanalytic perspective, it is always essential to ask what is the perspective of the person speaking in relation to their own speech acts? For instance, why does Damasio have to frame his discourse by inventing an imaginary friend? Who is this other to Damasio, and how does it relate to his own theory of mental subjectivity? Does he have to invent an imaginary other in order to convince himself that he is engaging in an open debate even when he is writing from a mono-logical perspective? Furthermore, in a book dedicated to showing how normal minds use reason to adapt to particular social situations in order to promote their own survival, why it is necessary for him to turn to his imaginary ideal friend?

The presence of this other within the self can be seen as disrupting Damasio's "rational" discourse from within. Although he mostly speaks from a perspective of absolute knowledge and self-consciousness, we have to ask how does this position of enunciation challenge his stress on non-conscious mental processes? As Lacan says in relation to Descartes, he may say "I think, therefore, I am," but he can only think this by writing it down.[55] There is thus a difference between what he is saying and how he says it, and with Damasio, we find this same type of division or splitting between what is argued and how it is argued.

Damasio relies on his reader to read his text as being objective, neutral, natural, and universal, but he himself does not allow for this type of reason to be accounted for under his category of reason. I see this conflict as being part of neoliberalism because as Lacan argues, the contemporary ego tries to get itself recognized by an idealized Other who is only able to verify the subject's own ideal ego. In this narcissistic transference, one speaks as an other to an Imaginary other as one represses the unconscious and the possible criticism of the social Other.[56] In this context, the psychoanalytic use of free association and the neutrality of the analyst represent the opposite of the idealized relationship between the self and the other. In analytic practice, the patient is told to suspend all self-censorship and to speak without intention, while the analyst is also supposed to refrain from judging or understanding. Just as Descartes allows himself to discover new things by following the paths of his discourse, psychoanalysis is centered on the creation of a new way of speaking and thinking.

The privileging of language through free association is clearly missing in Damasio's account of the human mind, and one reason for this lack is that he cannot help but return to the neoliberal focus on the individual who privatizes all social relationships:

> First, reaching a decision about the typical personal problem posed in a social environment, which is complex and whose outcome is uncertain, requires both broad-based knowledge and reasoning strategies to operate over such knowledge. The broad knowledge includes facts about objects, persons, and situations in the external world. But because personal and social decisions are inextricable from survival, the knowledge also includes facts and mechanisms concerning the regulation of the organism as a whole. The reasoning strategies revolve around goals, options for action, predictions of future out-come, and plans for implementation of goals at varied time scales. Second, the processes of emotion and feeling are part and parcel of the neural machinery for biological regulation, whose core is constituted by homeostatic controls, drives, and instincts. Third, because of the brain's design, the requisite broad-based knowledge depends on numerous systems located in relatively separate brain regions rather than in one region.[57]

Here every aspect of the human mind and social interaction is absorbed back into the individual's brain, which is driven by evolution to survive through adaptation. Even when Damasio does indicate the difference between images and words, he does it so that he can show how language can translate images into usable objects for mental reflection. There is no sense here that words can be overdetermined, poetic, nonsensical, social, or transcendent; words are simply tools to be used to master the self and the external world.

For Freud, the speech of free association helped him to discover that the ego represses the unconscious and that the unconscious is a system of memories placed in a network of associations. Through the interpretation of his dreams and the unconscious formations of others, Freud discovered that our minds automatically present past feelings and interpretations in a distorted form in order to avoid an internal censor.[58] Employing the poetic functions of substitution (metaphor) and displacement (metonymy), the unconscious is able to circumvent the censor and materialize past feelings through visualization and linguistic representation. Unfortunately, the new brain sciences of neuroscience, evolutionary psychology, and behavioral economics maintain a pre-psychoanalytic understanding of language, the unconscious, and human subjectivity. From this perspective, we can consider these new disciplines to be repressions of psychoanalysis and the unconscious itself.

NOTES

1. Damasio, Antonio R. *Descartes' error*. Random House, 2006.
2. Ibid., xiii.
3. Dover, Gabriel. "Anti-Dawkins." *Alas poor Darwin: Arguments against evolutionary psychology* (2000): 55–78.
4. Freud, Sigmund. "Formulations on the two principles of mental functioning." *The Standard Edition of the Complete Psychological Works of Sigmund Freud, Volume XII (1911–1913): The Case of Schreber, Papers on Technique and Other Works*. 1958. 213–226.
5. Freud, Sigmund. *Beyond the pleasure principle*. Vol. 840. Penguin UK, 2003.
6. We can interpret Freud's insistence on finding a biological cause for everything he could not explain as a symptom of his unanalyzed transference to science. In other words, Freud turns to an idealized vision of biology in order to resolve contradictions within his own non-biological theory. However, it is still vital to stress how Freud turns to biology to reveal all of the ways evolution fails to account for human behavior and thinking.
7. Damasio, 111.
8. Dawkins, Richard. *The selfish gene*. Oxford university press, 2016.

9. Sloterdijk, Peter. "Critique of cynical reason." (1988).
10. Damasio, 11.
11. Ibid., 111.
12. Ibid., xvi.
13. Lacan, Jacques. *The Four Fundamental Concepts of Psychoanalysis*, trans. Alan Sheridan." *New York: Norton* 67 (1978): 162–179.
14. Loose, Rik. *The subject of addiction: Psychoanalysis and the administration of enjoyment.* Karnac Books, 2002.
15. Freud, Sigmund. "The economic problem of masochism." *The Psychoanalytic Review (1913–1957)* 16 (1929): 209.
16. Damasio, 124.
17. Freud, Sigmund, and James Strachey. *The ego and the id.* No. 142. WW Norton & Company, 1962.
18. Freud, Sigmund. "Predisposition to the Obsessional Neurosis." *The Psychoanalytic Review (1913–1957)* 21 (1934): 347.
19. Ibid., 124.
20. Ibid., 125.
21. Lasch, Christopher. *The culture of narcissism: American life in an age of diminishing expectations.* WW Norton & Company, 1991: 63.
22. Ibid., 65–66.
23. Ibid., 66.
24. Ibid.
25. Ibid., 68.
26. Lewontin, Richard C., Steven Rose, and Leon J. Kamin. "Not in our genes: Biology, ideology, and human nature." (1984).
27. Gottlieb, Gilbert. "Probabilistic epigenesis." *Developmental Science* 10.1 (2007): 1–11.
28. Lacan, Jacques. "Remarque sur le rapport de Daniel Lagache." *Écrits, op. cit*: 647–684.
29. Klein, Naomi. *The shock doctrine: The rise of disaster capitalism.* Macmillan, 2007.
30. Damasio, 125.
31. Freud, Sigmund. "Psychopathology of everyday life." (1938).
32. Damasio, 126.
33. Ibid., 191.
34. Ibid., 229.
35. Gould, Stephen Jay. "Darwinian fundamentalism." *New York Review of Books* 44 (1997): 34–37.
36. Damasio, 247.
37. Descartes, 1.
38. Laclau, Ernesto, and Chantal Mouffe. *Hegemony and socialist strategy: Towards a radical democratic politics.* Verso, 2001.

39. Descartes, 26.
40. Žižek, Slavoj. *Looking awry: An introduction to Jacques Lacan through popular culture*. MIT press, 1992: 64.
41. Descartes, 26.
42. Ibid., 33.
43. Ibid., 36.
44. Lacan, Jacques. *The four fundamental concepts of psycho-analysis*. Vol. 11. WW Norton & Company, 1998: 36.
45. Damasio, 248.
46. Descartes, 46.
47. Lacan, *Four*, 35–37.
48. Damasio, 3–7.
49. Damasio, 15.
50. Damasio, 165.
51. Ibid., xvii.
52. James, Oliver. *They f*** you up: How to survive family life*. A&C Black, 2007.
53. Gould, Stephen Jay. "More things in heaven and earth." *Alas poor Darwin: Arguments against evolutionary psychology* (2000): 101–126.
54. Damasio, xviii.
55. Lacan, *Four*, 36.
56. Lacan, Jacques. "The subversion of the subject." *Ecrits, trans. A. Sheridan* (1967): 232–325.
57. Damasio, 83.
58. Freud, Sigmund. *The interpretation of dreams*. Read Books Ltd., 2013.

CHAPTER 3

The Backlash Politics of Evolutionary Psychology: Steven Pinker's *Blank Slate*

Abstract This chapter examines Steven Pinker's *Blank Slate* to demonstrate how the new brain sciences are often shaped by an academic effort to discredit the social sciences and replace them with a new form of social Darwinism. Moreover, this reaction against other academic disciplines mirrors an attempt to formulate a science-based conservative backlash against postmodern social movements, welfare state policies, and progressive parenting. We shall see that this political discourse is presented in the form of a value-free scientific theory, which itself attempts to deny the importance of culture, language, and history. In short, new brain sciences tend to posit that all other social science and humanities disciplines are not only misleading but also dangerous because they are not based on empirical facts. Just as psychoanalysis has been criticized for not being scientific, I will argue that evolutionary psychology attempts to eliminate the Freudian conceptions of the unconscious and sexuality in order to reimagine a new mode of neoliberal social Darwinism.

Keywords Steven Pinker • Neoliberalism • Evolutionary psychology • Backlash • Conservative • Social sciences

In the previous chapter, I examined how Damasio's vision for neuroscience is shaped by a neoliberal ideology that focuses on the way individuals are motivated to conform to an unequal social system by naturalizing socially constructed political systems. What I will show here is how the new brain sciences are at times shaped by an academic effort to discredit the social sciences and replace them with a new form of social Darwinism. Moreover, this reaction against other academic disciplines mirrors an attempt to formulate a science-based conservative backlash against post-modern social movements, welfare state policies, and progressive parenting. We shall see that this political discourse is presented in the form of a value-free scientific theory, which itself attempts to deny the importance of culture, language, and history.[1] In short, new brain sciences tend to posit that all other social science and humanities disciplines are not only misleading but also dangerous because they are not based on empirical facts. Just as psychoanalysis has been criticized for not being scientific, we shall see how evolutionary psychology attempts to eliminate the Freudian conceptions of the unconscious and sexuality in order to reimagine a new mode of neoliberal social Darwinism.[2]

Steven Pinker's *Blank Slate* is an important example of the current conservative scientific backlash discourse, and I turn to this text because it openly addresses the different discourses and politics it is attempting to replace. Although it rarely deals with psychoanalysis directly, Pinker's book exemplifies the repression of the unconscious and the dismissal of Freudian theory and practice. As a prime example of neoliberal ideology, the underlying argument is that we are determined through evolution and our genes to replicate behaviors and thoughts shaped mainly by our hunter-gatherer ancestors. This theory will be shown to be conservative in two ways: on one level, it argues that all we can do is conserve and repeat past social formations, and on another level, it feeds into the neoliberal conservative efforts to delegitimize the welfare state and progressive social movements. Ironically, at the same time Pinker spends hundreds of pages attacking progressive postmodern culture and politics, he denies the importance of culture in shaping human behavior.

Nature or Nurture Again

The first strategy Pinker uses to discredit the social sciences and progressive culture is to present an exaggerated version of the nature versus nurture debate. As a common ploy in conservative political rhetoric, Pinker begins by showing how scientists believing in nature and biology have been victimized by the liberal social sciences: "For invoking nurture and nature, not nurture alone, these authors have been picketed, shouted down, subjected to searing invective in the press, even denounced in Congress. Others expressing such opinions have been censored, assaulted, or threatened with criminal prosecution."³ Although we often think that the conservative, religious Right is behind the attack on the science of evolution, Pinker wants to argue that it is the liberal Left that refuses to take into account any role natural evolution plays in the formation of human subjectivity. We will see why he has to make this move, but what is essential to notice here is the role played by political ideology and rhetoric in his supposedly scientific discourse.

Like many other evolutionary psychologists, Pinker has to prove how his turn to evolution to explain mental traits is not a new form of social Darwinism; in other words, he wants to return to biological determinism without being connected to the history of using evolution to demonize different races and social groups.⁴ In what we can read as a return of the repressed, his own fear of being a scientific racist is both expressed and dismissed: "To acknowledge human nature, many think, is to endorse racism, sexism, war, greed, genocide, nihilism, reactionary politics, and neglect of children and the disadvantaged. Any claim that the mind has an innate organization strikes people not as a hypothesis that might be incorrect but as a thought it is immoral to think."⁵ The rhetorical strategy here is to present the critics of evolutionary psychology as being radicals who attack anyone who believes in the possibility of human nature. As a preemptive attack, Pinker presents the real possible criticism of his work only to dismiss them from a position of self-awareness. In other terms, he tries to defang his critics by showing that he is aware of their criticisms; of course, he also constructs the other side's arguments in an extreme way so that they are easy to dismiss, and yet, it is hard to ignore the possible racist and sexist implications of his own theory.

From a psychoanalytic perspective, we can understand Pinker's discourse as an Imaginary system that divides the world into a good "Us" and a bad "Them." Drawing from Lacan's theory of narcissism and paranoid knowledge, I argue that this splitting of the self off from the other is coupled with a desire to defend one's body of knowledge as if one is defending the integrity of one's own body.⁶ Lacan posits that since one first gains a sense of one's body as being complete and whole by identifying with an ideal image in the mirror or with a similar other, our sense of self is Imaginary and illusionary. Moreover, this ideal ego has to be defended and shapes our relationship with objects and our consciousness. As an Imaginary formation, the ego is not defined by material reality and allows for a high degree of mental autonomy, and it is this notion of the ego that is often repressed by evolutionary psychologists, neuroscientists, and behavioral economists.

In the case of Pinker, his efforts to defend his own knowledge shape how he differentiates evolutionary psychology from other academic disciplines. Since he must protect his knowledge at all costs, he is pushed to split off any idea that threatens his "body" of knowledge, and this split-off awareness is projected onto others, who he then attacks. We shall see that this combination of splitting, rejecting, projecting, and attacking is a key aspect to neoliberal conservative ideology and discourse. In Pinker's case, the rejected other of his argument is the liberal social scientist who believes that we are born as blank slates, and only culture, experience, and education shape our minds. While he traces the history of this idea of the blank slate, he represents his opponents in extreme ways as he claims the moderate middle ground for himself and the people who support his version of evolutionary psychology: "My goal in this book is not to argue that genes are everything and culture is nothing—no one believes that—but to explore why the extreme position (that culture is everything) is so often seen as moderate, and the moderate position is seen as extreme."[7] Once again, we have to look at the rhetorical and psychological construction of his argument, which functions by equating social scientists and humanists with the extreme idea that biology does not affect human nature; meanwhile, his own extreme statements are labeled as moderate. By stressing the pathological nature of his argument, I intend to show how the presentation of science is often shaped by the psychological and political manipulation of language. This type of analysis that I am presenting will require a careful attention to the words Pinker uses and the unfolding of his arguments and rhetoric. As a mode of psychoanalytic rhetorical criticism, I focus on the conscious and unconscious manipulation of language that often supports contemporary political discourse.

The Ideology of Non-ideology

An example of how Pinker's discourse is affected by a need to split off the repressed awareness of the limitations of his own beliefs can be seen by the fact that at the very moment that he makes highly charged political claims, he desires to represent his work as being non-ideological: "Nor does acknowledging human nature have the political implications so many fear. It does not, for example, require one to abandon feminism, or to accept current levels of inequality or violence, or to treat morality as a fiction. For the most part I will try not to advocate particular policies or to advance the agenda of the political left or right."[8] Although he wants to pretend that his discourse is apolitical, we shall see that he constantly rails against the Left and its supposed rejection of science, biology, evolution, and human nature. Moreover, while he argues that his work is not racist or sexist, his theory points to a naturalization of racist and sexist social hierarchies.

In the passage cited above, Pinker claims that he is not interested in advocating any particular politics, but he soon reveals that he does indeed seek to undermine a whole set of progressive public policies:

> First, the doctrine that the mind is a blank slate has distorted the study of human beings, and thus the public and private decisions that are guided by that research. Many policies on parenting, for example, are inspired by research that finds a correlation between the behavior of parents and the behavior of children. Loving parents have confident children, authoritative parents (neither too permissive nor too punitive) have well-behaved children, parents who talk to their children have children with better language skills, and so on. Everyone concludes that to grow the best children, parents must be loving, authoritative, and talkative, and if children don't turn out well it must be the parents' fault. But the conclusions depend on the belief that children are blank slates. Parents, remember, provide their children with genes, not just a home environment. The correlations between parents and children may be telling us only that the same genes that make adults loving, authoritative, and talkative make their children self-confident, well-behaved, and articulate.[9]

Once again, Pinker's strategy is to represent progressive beliefs in an extreme manner so that he can reject their theories and policies as he claims that he does not have a political agenda. In this case, the target is progressive parenting, which he derides for being based on the rejection of genetics and the affirmation of the utopian idea that people can be educated and taught because they are born without any biological predetermination.

On one level, Pinker is correct to critique the social scientists who dismiss genetics and biology in order to affirm that people are completely shaped by education, experience, and society; however, very few social scientists actually hold this extreme view, and Pinker tends to promote his own extreme position by returning to an older model of biological determinism. In other words, Pinker projects extremism onto his hated other so that he can deny his own extremism. Furthermore, it is hard to imagine how we can use public policy, education, and parenting to rewrite children's genetic inheritance, but it is possible to intervene to affect the mixture of genes, experience, personal psychology, education, and culture. Instead of accepting the middle ground and declaring that there is always a dialectic between nature and nurture, Pinker wants to hold onto the idea that we are shaped by a biologically determined universal human nature:

> The taboo on human nature has not just put blinkers on researchers but turned any discussion of it into a heresy that must be stamped out. Many writers are so desperate to discredit any suggestion of an innate human constitution that they have thrown logic and civility out the window. Elementary distinctions—"some" versus "all," "probable" versus "always," "is" versus "ought"—are eagerly flouted to paint human nature as an extremist doctrine and thereby steer readers away from it. The analysis of ideas is commonly replaced by political smears and personal attacks. This poisoning of the intellectual atmosphere has left us unequipped to analyze pressing issues about human nature just as new scientific discoveries are making them acute.[10]

Here we find a common paranoid Right-wing rhetorical strategy, which involves attacking others for what one is doing oneself.[11] Thus, the very moment that Pinker attacks researchers for attacking advocates of human nature, he pretends that he himself is not engaged in the very type of discourse he is critiquing. As he represents his adversary as being extremist, he uses an extreme account of their arguments. What, then, makes this discourse Right-wing is the way it positions the critic of progressive policies and beliefs as being the victims of liberal censorship.[12] Since Pinker believes that his own discourse is purely apolitical and scientific, he does not acknowledge the underlying political assumptions behind his arguments.[13]

From a psychoanalytic perspective, Pinker's rhetoric is determined by an unacknowledged use of splitting, mirroring, and projection: he first divides the world into those who believe in biologically determined human nature and those who do not, and then he attacks others for doing exactly what he is doing, which is making extreme and divisive claims in an effort to censor differing perspectives.[14] neoliberal conservatives often use this rhetorical psychological strategy in an unintentional way: they believe that they are right and good and that they have been persecuted by a liberal culture that does not agree with their beliefs. By positioning themselves as victims of liberal aggression and political correctness, they are able to justify their own aggression as they refuse any type of criticism.[15] Since the victim is always pure and good, you cannot attack the victim, and even when the victim lashes out, their aggression is justified and blameless. Pinker thus frames his "scientific" discourse through a political and psychological rhetoric that attempts to demonize anyone who does not agree with his extreme views. An obvious example of this political use of victim identity can be seen in Donald Trump, who also obsesses over how he is a victim of the liberal media and political correctness. While many on the Left argue that language plays an important role in defining identities and social relations, neoliberal conservatives tend to argue that the Left's attention to language use results in the formation of an oppressive superego that censors our true feelings and desires.

It is important to stress here that my psychoanalytic interpretation of Pinker's words and rhetorical strategies is necessary because underlying both the supposed neutrality of the new brain sciences and the ideology of conservative neoliberalism, we find a group of unconscious defense mechanisms that are not purely rational or intentional. In other words, one cannot fully understand Pinker's discourse and other conservative rhetoric, if one does not grasp the unconscious use of splitting, mirroring, and projection. Moreover, it is vital to also understand how conservatives position themselves to be victims of liberal culture, and this victimization relates to Freud's notion that people often fantasy about their own abuse because they find unconscious pleasure in suffering. Furthermore, as Lacan insists, people try to escape their own responsibility by blaming others, and since the victim cannot be criticized or attacked, the imagined victim identity helps to protect the victim's revenge from criticism or social censure.

Pinker's blindness to his own rhetoric and politics should not be too surprising because part of his mission is to discredit the belief in the ability of culture, subjectivity, and language to shape our inherited human nature:

I first had the idea of writing this book when I started a collection of astonishing claims from pundits and social critics about the malleability of the human psyche: that little boys quarrel and fight because they are encouraged to do so; that children enjoy sweets because their parents use them as a reward for eating vegetables; that teenagers get the idea to compete in looks and fashion from spelling bees and academic prizes; that men think the goal of sex is an orgasm because of the way they were socialized. The problem is not just that these claims are preposterous but that the writers did not acknowledge they were saying things that common sense might call into question. This is the mentality of a cult, in which fantastical beliefs are flaunted as proof of one's piety. That mentality cannot coexist with an esteem for the truth, and I believe it is responsible for some of the unfortunate trends in recent intellectual life. One trend is a stated contempt among many scholars for the concepts of truth, logic, and evidence. Another is a hypocritical divide between what intellectuals say in public and what they really believe. A third is the inevitable reaction: a culture of "politically incorrect" shock jocks who revel in anti-intellectualism and bigotry, emboldened by the knowledge that the intellectual establishment has forfeited claims to credibility in the eyes of the public.[16]

In response to this cultural criticism, one has to ask why someone who thinks cultural influence is so unimportant would spend so much time and energy attacking people who believe in cultural influence. Of course, Pinker does not believe he is doing cultural criticism; he believes he is doing science, and yet he exposes how science never is produced or communicated in a cultural vacuum. There is, however, a rhetoric, psychology, and politics to scientific discourse even if someone like Pinker wants to dismiss these influences.[17]

From Pinker's perspective, the unscientific nature of the social sciences has turned academic discourse into a cult based on fantastical beliefs, while the biological sciences provide all of the true answers about human nature and life. Due to the supposed loss of credibility for intellectual truth caused by the social sciences and progressive politics and policies, he argues a new politically incorrect culture has been produced. In other words, Pinker is blaming the anti-liberal discourse on liberals who have opened themselves up to criticism by not accepting biological determinism. For the narcissistic victim, the world is a mirror that reflects the subject's own aggression: therefore, as Pinker attacks others, he represents himself as being attacked.[18]

Although Pinker may consider himself to be a progressive or liberal, his underlying ideology argues that the liberal belief that human nature is not determined by nature results in totalitarian social planning. In other terms, people who believe that humans can be shaped by education, experience, and individual choice are really totalitarian Communists:

The belief that human tastes are reversible cultural preferences has led social planners to write off people's enjoyment of ornament, natural light, and human scale and force millions of people to live in drab cement boxes. The romantic notion that all evil is a product of society has justified the release of dangerous psychopaths who promptly murdered innocent people. And the conviction that humanity could be reshaped by massive social engineering projects led to some of the greatest atrocities in history.[19]

For neoliberal conservatives, the enemy is often "social planning" because it goes against the free market and individual freedom, and it often requires some redistribution of wealth through taxation.[20] By equating critics of neurobiology with totalitarian governmental control, Pinker is able to promote a libertarian politics at the same time he discredits the modern welfare state.[21] Instead of arguing that wealthy people should not have to pay taxes to support expensive welfare programs for the poor, he argues that the belief in state-sponsored welfare programs is based on a denial of the science of human nature. From this extreme perspective, progressive parenting and liberal politics lead to the atrocities of totalitarianism.

After expressing pages of heated rhetoric, Pinker has the audacity to represent his discourse as being "coolly analytical" and scientific:

> Though many of my arguments will be coolly analytical—that an acknowledgment of human nature does not, logically speaking, imply the negative outcomes so many people fear—I will not try to hide my belief that they have a positive thrust as well. "Man will become better when you show him what he is like," wrote Chekhov, and so the new sciences of human nature can help lead the way to a realistic, biologically informed humanism. They expose the psychological unity of our species beneath the superficial differences of physical appearance and parochial culture.[22]

The key then to Pinker's "realistic, biologically informed humanism" is his belief in a universal human nature that denies the superficial differences of "parochial culture," and this denial of cultural influences and social difference is required by a theory that wants to posit that there is a universal human nature inherited through evolution.[23] Therefore, as much as Pinker wants to deny that he is an extremist in his biological determinism, his own text reveals a deep commitment to the genetic control of mental functions and attitudes.

Like so many other evolutionary psychologists, Pinker argues that his science is a critique of social theory and does not need theory since it is a pure science[24]: "They [theories of human nature] promise a naturalness in human relationships, encouraging us to treat people in terms of how they do feel rather than how some theory says they ought to feel."[25] Evolutionary psychology thus attempts to naturalize culture and social relations as it presents a theory that denies that it is a theory. As an anti-ideological ideology, Pinker's discourse reveals the hidden truth about neuroliberalism: the new brain sciences are often a political rhetoric disguised as natural facts, and this naturalization of social constructions serves to rationalize the neoliberal political and social status quo.[26]

The Blank Slate of Democracy

As I argued in the previous chapter, neuroscience has a difficult time dealing with abstract principles like democracy. Moreover, I claimed that in Descartes' discourse, we see how modernity begins with a leap into reason and democracy, and this leap can be seen as a break with biology, evolution, and tradition. However, for Pinker, this modern call for equality and universal reason is problematic because it relies on the anti-evolutionary notion of the blank slate:

> Locke's notion of a blank slate also undermined a hereditary royalty and aristocracy, whose members could claim no innate wisdom or merit if their minds had started out as blank as everyone else's. It also spoke against the institution of slavery, because slaves could no longer be thought of as innately inferior or subservient.[27]

Pinker's rejection of Locke's use of the blank slate here begs the question of whether Pinker endorses slavery or hereditary royalty: after all, he argues that the modern rejection of these premodern institutions was based on the false idea that humans are shaped by their experiences and histories and not some inherited force, be it genes or aristocratic inheritance. If one does believe in the determining power of biology, how does one not escape from arguing that slaves are slaves because they have inherited inferior genes?

The modern democratic solution to both biological and social inheritance is to socially construct institutions and beliefs dedicated to universal human rights and justice. These social formations are not derived from nature, and they may in fact be counter to many natural tendencies, but to dismiss them is to reject the very possibility of a more equal and just society. Unfortunately, Pinker does not recognize the necessity of this leap into democracy, and instead, he attacks the proponents of the blank slate doctrine:

> During the past century the doctrine of the Blank Slate has set the agenda for much of the social sciences and humanities. As we shall see, psychology has sought to explain all thought, feeling, and behavior with a few simple mechanisms of learning. The social sciences have sought to explain all customs and social arrangements as a product of the socialization of children by the surrounding culture: a system of words, images, stereotypes, role models, and contingencies of reward and punishment ... According to the doctrine, any differences we see among races, ethnic groups, sexes, and individuals come not from differences in their innate constitution but from differences in their experiences. Change the experiences—by reforming parenting, education, the media, and social rewards—and you can change the person. Underachievement, poverty, and antisocial behavior can be ameliorated; indeed, it is irresponsible not to do so. And discrimination on the basis of purportedly inborn traits of a sex or ethnic group is simply irrational.[28]

Here we find a prime example of the conservative side of neuroliberalism: from this anti-democratic perspective, since our universal human nature is determined by our genes selected through evolution, it is absurd to believe that society is socially constructed and that we are born into the world open to being shaped by cultural and personal experiences.[29] Pinker takes a strong stand here against 300 years of social progress as he appears to endorse racism, sexism, and ethnocentrism. From his perspective, it is irrational to try to reverse poverty, underachievement, or anti-social behavior because these issues are caused by genetic factors.

My argument here is not that Pinker is intentionally trying to use his view of science as a weapon against progressive public policies; rather, the paranoid ideological construction of science that he endorses functions to undermine any political and social intervention relying on the ability of people to overcome their inherited biology.[30] Moreover, psychoanalysis can play a key role here in providing a counter-discourse to Pinker's conservative neuroliberalism since psychoanalysis shows the many ways that human beings go against nature, instincts, evolution, self-regulation, and biological determinism. Although psychoanalysis recognizes the power of biology in shaping human nature, it also posits that inherited instincts are disrupted by subjectivity, language, and culture. From this perspective, we are not born with a blank slate, but we also are not born predetermined by nature.

AGAINST THE PRIMITIVE

As Pinker rejects the notion of the blank slate, he also rejects what he sees as the liberal idealization of "primitive" cultures: "The concept of the noble savage was inspired by European colonists' discovery of indigenous peoples in the Americas, Africa, and (later) Oceania. It captures the belief that humans in their natural state are selfless, peaceable, and untroubled, and that blights such as greed, anxiety, and violence are the products of civilization."[31] The conservative agenda behind this rejection of the Noble Savage is to show how humans are naturally aggressive, selfish, troubled, greedy, anxious, and violent, and so progressive policies, parenting, and education will always fail because they do not accept the dark side of universal human nature: "No one can fail to recognize the influence of the doctrine of the Noble Savage in contemporary consciousness. We see it in the current respect for all things natural (natural foods, natural medicines, natural childbirth) and the distrust of the man-made, the unfashionability of authoritarian styles of childrearing and education, and the understanding of social problems as repairable defects in our institutions rather than as tragedies inherent to the human condition."[32] Once again, science is here used to discredit the idea of making society more just and fair because "human defects" are the product of nature and not culture, and one of the results of this discourse is to enforce a neoliberal justification for eliminating the welfare state and allowing the free reign of the natural free market.

Like so many other neoliberal backlash conservatives, Pinker promotes the idea that individual ethnic and racial groups are endowed with different genetic material as he attempts to deny the fact that discrimination and racism are still prevalent in our contemporary world:

> Today no respectable public figure in the United States, Britain, or Western Europe can casually insult women or sling around invidious stereotypes of other races or ethnic groups. Educated people try to be conscious of their hidden prejudices and to measure them against the facts and against the sensibilities of others. In public life we try to judge people as individuals, not as specimens of a sex or ethnic group. We try to distinguish might from right and our parochial tastes from objective merit, and therefore respect cultures that are different or poorer than ours. We realize that no mandarin is wise enough to be entrusted with directing the evolution of the species, and that it is wrong in any case for the government to interfere with such a personal decision as having a child. The very idea that the members of an ethnic group should be persecuted because of their biology fills us with revulsion.[33]

Here it appears that Pinker is taking a stand against racism and sexism, but his need to attack social progressives and politically correct social scientists forces him to return to a form of social Darwinism that can only result in naturalizing destructive discourses of racial and sexual prejudice[34]:

> Academics were swept along by the changing attitudes to race and sex, but they also helped to direct the tide by holding forth on human nature in books and magazines and by lending their expertise to government agencies. The prevailing theories of mind were refashioned to make racism and sexism as untenable as possible. The doctrine of the Blank Slate became entrenched in intellectual life in a form that has been called the Standard Social Science Model or social constructionism. The model is now second nature to people and few are aware of the history behind it.[35]

In his efforts to promote nature over nature, Pinker rejects what evolutionary psychologists call the "Standard Social Science Model" and the idea that many of our ideas and beliefs are socially constructed.[36] One possible effect of this rhetoric is to discredit other academic disciplines so that the new brain sciences can gain institutional prestige, which will lead to more funding and power. However, another result of this discourse is to counter welfare state policies and liberal culture by discrediting the social science theories and research that legitimate social intervention.

As we found with the case of the concept of democracy, social scientists and humanists have often promoted the impossible, but necessary, ideals of universal equality and justice not because they think these cultural goals are our natural tendency but because they want to see a more just form of society.[37] However, for Pinker, and many other evolutionary psychologists, the progressive attempts to fight racism and sexism can only fail because liberals do not recognize the biological foundation of human nature. From Pinker's perspective, social policy should be based on science, and the science of human nature should be based on evolution, and any theory or policy that relies on the social construction of values and other mental categories will be counter-productive. In taking a strong but subtle stand against equality and justice, Pinker uses science to justify oppression at the same time he attacks people for accusing him of what he is actually doing:

> Many of the pressing social problems of the first decades of the twentieth century concerned the less fortunate members of these groups. Should more immigrants be let in, and if so, from which countries? Once here, should they be encouraged to assimilate, and if so, how? Should women be given equal political rights and economic opportunities? Should blacks and

whites be integrated? ... These social challenges were not going to go away, and the most humane assumption was that all human beings had an equal potential to prosper if they were given the right upbringing and opportunities. Many social scientists saw it as their job to reinforce that assumption.[38]

The main idea that upsets Pinker here is the notion that all human beings have the equal potential to prosper. As we saw with Descartes, this idea of universal potentiality is at the heart of the modern quest for justice and equality, and yet Pinker finds this democratic impulse to be the recent invention of liberal social scientists who refuse to accept the truth about human nature. Like so many backlash conservatives, Pinker wants to deny the value of important social programs because he thinks that they lack a scientific basis, and they are supported by the misguided notion that people can be helped or educated. There is thus a fundamentally nihilistic belief behind this political use of scientific discourse.

As Pinker denies he has a political agenda, he continues to present social arguments that can only be considered to be regressive and discriminatory: "John Stuart Mill (1806–1873) was perhaps the first to apply his blank-slate psychology to political concerns we recognize today. He was an early supporter of women's suffrage, compulsory education, and the improvement of the conditions of the lower classes."[39] Since Pinker has already established that he thinks the theory of the blank slate advocated by Mill is wrong and dangerous, is there any way to read this argument other than the fact that Pinker is discrediting the right of women to vote, the need for universal education, and the desire to help the poor? Of course, Pinker would respond to this criticism by claiming that he is being victimized by people who do not understand science and who mistakenly endorse the false idea that society can affect basic mental human attributes.

One of the main targets of Pinker's criticism is the anthropologist Franz Boas who stressed the ways language and culture contribute to the mental differences among people:

Idealism allowed Boas to lay a new intellectual foundation for egalitarianism. The differences among human races and ethnic groups, he proposed, come not from their physical constitution but from their culture, a system of ideas and values spread by language and other forms of social behavior. Peoples differ because their cultures differ ... The idea that minds are shaped by culture served as a bulwark against racism and was the theory one ought to prefer on moral grounds. Boas wrote, "I claim that, unless the contrary can be proved, we must assume that all complex activities are socially determined, not hereditary."[40]

Pinker is perhaps correct in rejecting Boas' overemphasis on nurture over nature, but Pinker goes too far in simply reversing the opposition and rejecting any role for culture and language. Furthermore, what appears to really bother Pinker is Boas' idealism and his efforts to overcome racism through social intervention.

For a biological determinist like Pinker, social scientists who seek to reject all aspects of evolution and individual psychology in order to focus on the purely cultural aspects of human subjectivity are easy targets. As I stressed in the previous chapter, the new brain sciences tend to dismiss anything that cannot be measured in the brain scan of an isolated individual, and so any sense that social relations transcend the individual mind and brain has to be rejected. In Pinker's case, he rejects the possibility of any level of social autonomy freed from evolutionary control:

> The doctrine of the superorganism has had an impact on modern life that extends well beyond the writings of social scientists. It underlies the tendency to reify "society" as a moral agent that can be blamed for sins as if it were a person. It drives identity politics, in which civil rights and political perquisites are allocated to groups rather than to individuals. And as we shall see in later chapters, it defined some of the great divides between major political systems in the twentieth century.[41]

Following a common move of the neoliberal conservative backlash, Pinker attacks the notion of group values and identity politics: for the evolutionary psychologist, society is made out of isolated individuals determined by biology and not culture, politics, or language. Like so many other conservative thinkers, Pinker critiques the civil rights movement because it focused on the need to help a particular group and not isolated individuals. Following Margaret Thatcher's neoliberal claim that there is no society, Pinker tends to dismiss the very idea of the social transcending the individual, and this rejection of the social undermines the modern welfare state and any belief in collective action.[42]

Of course, the most conservative aspect of this form of neuroliberalism is the rejection of the ability of people or societies to change at all: "More generally, social scientists saw the malleability of humans and the autonomy of culture as doctrines that might bring about the age-old dream of perfecting mankind. We are not stuck with what we don't like about our current predicament, they argued. Nothing prevents us from changing it except a lack of will and the benighted belief that we are permanently consigned to it by biology." This rejection of progressive social change has been a core part of conservative ideology for centuries, but what has changed recently is that this fight against the promoters of progress has taken on a scientific foundation. While in the past, conservative thinkers like Hobbes and Burke referred to a vague sense of corrupted human nature to explain why radical, progressive social change cannot happen, the neuroliberal conservatives turn to evolution and brain images to prove the force of biological inheritance.[43]

One way that science is able to exert such a strong influence on society and culture is by denying its own political relevance. For instance, Pinker reveals this attempt to repress the political forces behind his discourse by blaming his opponents for being shaped by politics: "Though psychology is not as politicized as some of the other social sciences, it too is sometimes driven by a utopian vision in which changes in childrearing and education will ameliorate social pathologies and improve human welfare."[44] From Pinker's paranoid perspective, social scientists try to impose a utopian political agenda based on the idea that people can be transformed through experience, education, parenting, and politics, but his own discourse is purely scientific and stands above political manipulation. The truth of the matter is that all academic discourses are shaped by politics, culture, history, and ideology, and while Pinker tries to deny these influences, he continues to attack all progressive aspects of the modern welfare state.

Pinker reveals how his conservative neuroliberal rejection of the social science, social constructivism, and progressive politics centers on the notion that the social order has no independence from evolution and biology:

> Kroeber wrote: "The dawn of the social ... is not a link in any chain, not a step in a path, but a leap to another plane.... [It is like] the first occurrence of life in the hitherto lifeless universe.... From this moment on there should be two worlds in place of one." And Lowie insisted that it was "not mysticism, but sound scientific method" to say that culture was "sui generis" and could be explained only by culture, because everyone knows that in biology a living cell can come only from another living cell.[45]

Once again, Pinker invokes here an extreme example of social determinism in order to undermine the role played by culture and language in shaping human subjectivity and society. It should be no wonder then that the new brain sciences end up reinforcing the neoliberal privileging of the individual over the social state since evolutionary psychology tends to reject the possibility that social relations can transcend individual brains.

It should be noted here that one of the difficulties dealing with the paranoid rhetoric of the Right is that one cannot help but fall into oppositional thinking since one is critiquing a system of thought that continues to divide everything in a binary manner. Thus, Pinker could argue that my critique of his work is also based on a division between the good liberal social scientists and the bad conservative evolutionary psychologists; however, what I am trying to show is that psychoanalysis offers us a way of thinking how nature and nurture interact, and from this perspective, there is a constant dialectical interplay between biology and society mediated by psychology and the unconscious.

As we see throughout his work, Pinker often makes an extreme claim about evolutionary determinism, and then he defends his argument by showing that he is not making an extreme claim. For instance, in the following passage, he points to a more dialectical interaction between nature and culture:

> We might all be equipped with a program that responds to an affront to our interests or our dignity with an unpleasant burning feeling that motivates us to punish or to exact compensation. But what counts as an affront, whether we feel it is permissible to glower in a particular setting, and what kinds of retribution we think we are entitled to, depend on our culture. The stimuli and responses may differ, but the mental states are the same, whether or not they are perfectly labeled by words in our language.[46]

Here he posits that a particular culture might affect what offends us and how we respond to an offense, but the idea of being offended is itself universal. However, it is unclear what he means when he says that the "mental states are the same" in different cultures. If he is claiming that we all have some shared mental response mechanisms but the input and the output differ, then the way culture affects the inputs and the outputs would still remain essential.

Evolutionary psychologists tend to make a series of contradictory arguments because they argue for a shared, universal, biologically determined human nature as they recognize the existence of cultural differences. The trick then is how to point to culture and deny its importance at the same time. In fact, Pinker's entire book is framed by this contradiction since he wants to reduce the value of social influences as he attacks the social influence of misguided social scientists. For instance, the following quote does appear to present a more balanced view, and yet, his own argument cannot allow for any real role for cultural determinism:

> The moral, then, is that familiar categories of behavior—marriage customs, food taboos, folk superstitions, and so on—certainly do vary across cultures and have to be learned, but the deeper mechanisms of mental computation that generate them may be universal and innate. People may dress differently, but they may all strive to flaunt their status via their appearance.[47]

On one level, Pinker is arguing something very basic, which is that all people use language and have similar emotional responses, but each culture affects when and how these emotions are presented and how language is used and experienced. Yet, to make his evolutionary argument, he has to claim that particular cultural representations and interactions are only superficial manifestations of universal mental processes.

The Neuroliberal Brain

An important argument for evolutionary psychology and neuroscience is to posit that biological universality is proven by the fact that all mental experiences have to pass through the evolved human brain: "One can say that the information-processing activity of the brain causes the mind, or one can say that it is the mind, but in either case the evidence is overwhelming that every aspect of our mental lives depends entirely on physiological events in the tissues of the brain."[48] From this perspective, no aspect of social interaction or culture can transcend the brain of an individual: in other words, culture can only develop and function by being processed by individual thinking. At first, this claim appears to be undeniable, but it ends up denying the fact that the whole of a group's activities can be greater than the sum of the individual contributions. Furthermore, since people can participate in cultural social formations that they do not understand or acknowledge, it must be possible that parts of the social go beyond the neural networks of discrete individuals. In other words, something like society or what Lacan calls the Other does affect our lives even if we do not know it.[49] However, from a conservative neoliberal perspective, it is essential to deny the importance of society and collective action because the focus is on the free individual participating in the unregulated free market without government intervention. In fact, libertarian conservatives often view the free market as an extension of evolutionary natural selection since the market is seen as an automatic system that picks winners and losers through free interaction. Just as supply automatically matches up with demand and prices spontaneously find the proper equilibrium, the neoliberal ideology of the free market is based on a naturalization of the social interaction. Since the neoliberal market is imagined to be both mechanical and natural, nature and culture are both redefined as being automatic.

For Pinker, if every human thought and emotion is the result of a universal mental mechanism, then it is possible for a machine to read our thoughts and feelings:

> Every emotion and thought gives off physical signals, and the new technologies for detecting them are so accurate that they can literally read a person's mind and tell a cognitive neuroscientist whether the person is imagining a face or a place. Neuroscientists can knock a gene out of a mouse (a gene also found in humans) and prevent the mouse from learning, or insert extra copies and make the mouse learn faster.[50]

Here we see how neuroscience and evolutionary psychology lead to the ultimate goal of connecting all of our mental experiences to particular genes that can be later manipulated through surgery or drug treatment. As I will discuss in Chap. 6, all of the new brain sciences are knowingly and unknowingly participating in a neuroliberal ideology that replaces the need for social change and political intervention with the promotion of particular drugs to deal with specific mental dysfunctions derived from evolution and genetics.

Part of this neuroliberal ideology involves not only denying the importance of the social and the cultural but also eliminating the value of the individual:

> Cognitive neuroscientists have not only exorcised the ghost but have shown that the brain does not even have a part that does exactly what the ghost is supposed to do: review all the facts and make a decision for the rest of the

brain to carry out. Each of us feels that there is a single "I" in control. But that is an illusion that the brain works hard to produce...[51]

Pinker's point is that the mechanical brain preprogrammed by evolution produces the illusion of the self and free will. Here we find an interesting reflection of neoliberal ideology, which often posits that we are free when we are being controlled, and we are controlled when we are free.[52]

Losing the Self and the Unconscious

At the very moment Pinker points to the illusion of free will and consciousness, he also returns to the notion that most mental functions and illnesses can be given an entirely genetic foundation:

> Autism, dyslexia, language delay, language impairment, learning disability, left-handedness, major depressions, bipolar illness, obsessive-compulsive disorder, sexual orientation, and many other conditions run in families, are more concordant in identical than in fraternal twins, are better predicted by people's biological relatives than by their adoptive relatives, and are poorly predicted by any measurable feature of the environment.[53]

By stressing biological determinism for psychological disorders, Pinker is able to clear a path for the elimination of psychoanalysis as a form of therapy, while the only option becomes a pharmaceutical solution. Here the goal of erasing both the social sciences and psychoanalysis is accomplished: since all social and mental problems are derived from biology, only biology can provide a solution. As I will argue in Chap. 6, this turn toward the evolutionary explanation for mental functioning and dysfunction feeds into the growth of the Governmental University Medical Pharmaceutical Complex because once genetics are seen as the controlling factor, the only solution is medication or direct medial intervention.

One of the key ways that Pinker "proves" this neurobiological foundation for psychological problems is through twin studies:

> Identical twins think and feel in such similar ways that they sometimes suspect they are linked by telepathy. When separated at birth and reunited as adults, they say they feel they have known each other all their lives. Testing confirms that identical twins, whether separated at birth or not, are eerily alike (though far from identical) in just about any trait one can measure. They are similar in verbal, mathematical, and general intelligence, in their

degree of life satisfaction, and in personality traits such as introversion, agreeableness, neuroticism, conscientiousness, and openness to experience.[54]

Although it would be wrong to dismiss or downplay the role of genes in contributing to some aspects of personality, Pinker wants to use these studies of identical twins to describe how many if not most of our mental characteristics are determined by genetics and genetics are determined by evolution.

Of course, identical twins represent extreme cases of sharing genetic material, and even in these cases, biology only plays a partial role in shaping mental characteristics:

> People sometimes fear that if the genes affect the mind at all they must determine it in every detail. That is wrong, for two reasons. The first is that most effects of genes are probabilistic. If one identical twin has a trait, there is usually no more than an even chance that the other will have it, despite their having a complete genome in common. Behavioral geneticists estimate that only about half of the variation in most psychological traits within a given environment correlates with the genes.[55]

This statement by Pinker reflects a much more balanced approach, but it calls into question much of what he has written in his book. If at most genes only determine half of a mental attribute, why does Pinker seem to dismiss non-biological factors? Also, why in his discussion of identical twins does he focus on how they express so many identical psychological features? Furthermore, if we can do very little about changing our genes, but we can try to change our social environments and individual responses, why does he seek to reject the need to change society or individual psychology?

I have been arguing that the main reason Pinker attacks non-biological theories is that his science is in actuality a political ideology, and his politics reinforces the neoliberal undermining of liberal culture and the postmodern welfare state. By arguing against the role played by nurture in shaping human nature, he is able to delegitimize educational, social, and governmental interventions into human suffering and inequality, yet one of the only alternatives left is to turn to drugs as the way to counteract inherited mental programs.

NOTES

1. Tallis, Raymond. *Aping mankind*. Routledge, 2016.
2. Lewontin, Richard C., Steven Rose, and Leon J. Kamin. "Not in our genes: Biology, ideology, and human nature." (1984).
3. Pinker, Steven. *The blank slate: The modern denial of human nature*. Penguin, 2003: viii.
4. Degler, Carl N. *In search of human nature: The decline and revival of Darwinism in American social thought*. Oxford University Press on Demand, 1991.
5. Pinker, viii.
6. Mills, Jon. "Lacan on paranoiac knowledge." *Psychoanalytic Psychology* 20.1 (2003): 30.
7. Ibid., ix.
8. Ibid.
9. Ibid.
10. Ibid., x.
11. John K. Wilson. *The myth of political correctness: The conservative attack on higher education*. Duke University Press, 1995.
12. Reisigl, Martin. *Analyzing political rhetoric*. Basingstoke: Palgrave, 2008.
13. Rose, Steven, and Hilary Rose. "Social Responsibility (III): The Myth of the Neutrality of Science." *Impact of Science on Society* 21.2 (1971): 137–149.
14. Samuels, Robert. "New media, cultural studies, and critical theory after postmodernism: Automodernity from Zizek to Laclau." (2010).
15. Cole, Alyson Manda. *The cult of true victimhood: from the war on welfare to the war on terror*. Stanford University Press, 2007.
16. Pinker, x.
17. Stewart, Larry. "The rise of public science: rhetoric, technology, and natural philosophy in Newtonian Britain, 1660–1750." (1992).
18. Kohut, Heinz. *The analysis of the self: A systematic approach to the psychoanalytic treatment of narcissistic personality disorders*. University of Chicago Press, 2013.
19. Pinker, x.
20. Ashbee, Edward. "Neoliberalism, conservative politics, and 'social recapitalization'." *Global Discourse* 5.1 (2015): 96–113.
21. Mendes, Philip. "Australian neoliberal think tanks and the backlash against the welfare state." *The Journal of Australian Political Economy* 51 (2003): 29.
22. Pinker, xi.
23. Cooley, Charles Horton. *Human nature and the social order*. Transaction Publishers, 1992.

24. Barkow, Jerome H., Leda Cosmides, and John Tooby, eds. *The adapted mind: Evolutionary psychology and the generation of culture.* Oxford University Press, 1995.
25. Pinker, xi.
26. Freeden, Michael. *Ideology: A very short introduction.* Vol. 95. Oxford University Press, 2003.
27. Pinker, 6.
28. Ibid.
29. For a critique of the nature versus nurture argument, see Shepherdson, Charles. *Vital signs: Nature, culture, psychoanalysis.* Psychology Press, 2000.
30. Hofstadter, Richard. *The paranoid style in American politics.* Vintage, 2012.
31. Pinker, 6.
32. Pinker, 8.
33. Pinker, 16.
34. Tileagă, Cristian. "Representing the 'Other': A discursive analysis of prejudice and moral exclusion in talk about Romanies." *Journal of Community & Applied Social Psychology* 16.1 (2006): 19–41.
35. Pinker, 16.
36. Cosmides, Leda, and John Tooby. "Cognitive adaptations for social exchange." *The adapted mind* (1992): 163–228.
37. Laclau, Ernesto, and Chantal Mouffe. *Hegemony and socialist strategy: Towards a radical democratic politics.* Verso, 2001.
38. Ibid.
39. Ibid., 18.
40. Ibid., 22.
41. Ibid., 26.
42. Prasad, Monica. *The politics of free markets: The rise of neoliberal economic policies in Britain, France, Germany, and the United States.* Vol. 19. Chicago: University of Chicago Press, 2006.
43. Kirk, Russell. *The conservative mind: from Burke to Eliot.* Regnery Publishing, 2001.
44. Pinker, 28.
45. Ibid.
46. Pinker, 39.
47. Ibid.
48. Ibid., 41.
49. Zizek, Slavoj. *On belief.* Routledge, 2003.
50. Pinker, 41.
51. Ibid., 42.

52. Samuels, Robert. "Auto-modernity after postmodernism: Autonomy and automation in culture, technology, and education." *Digital youth, innovation, and the unexpected* (2008).
53. Pinker, 46.
54. Pinker, 375.
55. Pinker, 376.

CHAPTER 4

Behavioral Economics: *Nudge* and Technocratic Liberalism

Abstract In this chapter, I examine the representation of behavioral economics in Cass Sunstein's and Richard Thaler's *Nudge*. In what they call "paternal libertarianism," we will find the foundations of a technocratic moderate middle ground, which combines together elements of liberal neuroscience and conservative evolutionary psychology. What all three of these approaches share is that the economic and social status quo is naturalized, and the result is what we can call a neuroliberal libertarian consensus. In the contemporary context, libertarianism refers to the belief in an unregulated free market that leads to the freedom of the individual. From this perspective, markets act as natural systems that choose winners and losers in an evolutionary survival of the fittest, and so any intervention into the market is seen as undermining the natural order. Moreover, libertarian economists, like Hayek, argue that since humans never have enough knowledge as individuals, state planning should be limited, and any attempt at social control could result in tyranny. Although Thaler and Sunstein are not pure libertarians, they seek to find a middle ground between conservative and liberal ideology.

Keywords Cass Sunstein • Richard Thaler • Libertarianism • Third Way • Behavioral economics

© The Author(s) 2017
R. Samuels, *Psychoanalyzing the Politics of the New Brain Sciences*,
https://doi.org/10.1007/978-3-319-71891-0_4

In the second chapter, I explored how Damasio's conception of neuroscience leads to a liberal form of neuroliberalism where humans are represented as being directed by evolution to pursue their own survival through conformity to particular social environments. Following this liberal version, I examined how Pinker's evolutionary psychology represents a conservative backlash version of neuroliberalism because the focus was on attacking progressive social sciences, welfare state policies, and liberal parenting. In this chapter, I examine the representation of behavioral economics in Cass Sunstein and Richard Thaler's *Nudge*.[1] In what they call "paternal libertarianism," we will find the foundations of a technocratic moderate middle ground, which combines together elements of liberal neuroscience and conservative evolutionary psychology. What all three of these approaches share is that the economic and social status quo is naturalized, and the result is what we can call a neuroliberal libertarian consensus.

THE LIBERTARIAN CONSENSUS

In the contemporary context, libertarianism refers to the belief in an unregulated free market that leads to the freedom of the individual.[2] From this perspective, markets act as natural systems that choose winners and losers in an evolutionary survival of the fittest, and so any intervention into the market is seen as undermining the natural order.[3] Moreover, libertarian economists, like Hayek, argue that since humans never have enough knowledge as individuals, state planning should be limited, and any attempt at social control could result in tyranny.[4] Although Thaler and Sunstein are not pure libertarians, they seek to find a middle ground between conservative and liberal ideology:

> The libertarian aspect of our strategies lies in the straightforward insistence that, in general, people should be free to do what they like—and to opt out of undesirable arrangements if they want to do so. To borrow a phrase from the late Milton Friedman, libertarian paternalists urge that people should be "free to choose." We strive to design policies that maintain or increase freedom of choice. When we use the term libertarian to modify the word paternalism, we simply mean liberty-preserving. And when we say liberty-preserving, we really mean it. Libertarian paternalists want to make it easy for people to go their own way; they do not want to burden those who want to exercise their freedom.[5]

Although libertarianism has often been associated with the conservative backlash against communism and liberal social welfare state policies, Thaler and Sunstein place this political philosophy within a more moderate or middle-ground discourse. They want people to be free to choose, but they also want the government to be able to nudge people in certain directions by designing particular economic "choice architectures."

Central to the libertarian foundation of behavioral economics is the notion that people often make the wrong decisions, and so they need a little help to steer clear of risky behavior: "Drawing on some well-established findings in social science, we show that in many cases, individuals make pretty bad decisions—decisions they would not have made if they had paid full attention and possessed complete information, unlimited cognitive abilities, and complete self-control."[6] Here we see how a central theme of behavioral economics is the notion that individuals tend to make bad choices because their own psychology and lack of knowledge get in the way. These economists, then, turn to cognitive psychology to reveal the mistakes that individuals often make as they demonstrate that one of the main reasons people are unable to act like rational actors is that they are dominated by unconscious processes.

From a certain perspective, the focus on the unconscious in behavioral economics would appear to open the door for a psychoanalytic contribution to the field, but we soon find out that psychoanalysis itself is repressed as the unconscious is simply equated with mental processes that are not conscious. Instead of considering the Freudian notion that unconscious processes are the result of repression and the desire to avoid thinking about amoral and disturbing subject matter, the tendency in behavioral economics is to turn to neuroscience in order to articulate intuitive and automatic mental processes, and here we see how they are able to avoid dealing with both the psychoanalytic unconscious and the Freudian notion of sexuality. Rather than centering self-destructive behavior on the sexualization of pain and punishment, behavioral economists and neuroscientists tend to connect self-defeating behavior to faulty logic and intuitive mental programs inherited through natural selection.

One way of explaining this repression of psychoanalysis is to recognize that since Freud has been attacked for not being scientific and empirical, the new brain sciences have sought to redefine the unconscious through neurobiology. Another reason for this move against psychoanalysis is that it opens up the door to test unconscious processes in a lab because once a specific mental function is tied to a particular brain region, it can be detected through scanning technology. However, at the same time these behavioral economists turn to neuroscientists to make their findings appear to be rational, objective, neutral, and universal, they continue to frame their assumptions by implicit and explicit political perspectives. For example, in the following description of their ideology, they display a conscious awareness of their own political frame:

> Libertarian paternalism is a relatively weak, soft, and nonintrusive type of paternalism because choices are not blocked, fenced off, or significantly burdened. If people want to smoke cigarettes, to eat a lot of candy, to choose an unsuitable health care plan, or to fail to save for retirement, libertarian paternalists will not force them to do otherwise—or even make things hard for them. Still, the approach we recommend does count as paternalistic, because private and public choice architects are not merely trying to track or to implement people's anticipated choices.[7]

The goal here appears to be to represent their paternal libertarianism as a middle ground that steers a "Third Way" between the pure libertarianism of the Right and the welfare state government of the Left; however, we have to realize that this middle ground is not neutral or unbiased since it represents a very particular perspective.[8]

This attempt to maintain a discourse of scientific truth at the same time that one points to a particular ideology is a key part of neuroliberalism where we often find a desire to combine ideological perspectives with a claim to be non-ideological. In this case, the repression of politics is centered on seeing the middle ground as being apolitical and neutral.[9] The idea is that if both liberal and conservative perspectives are combined, then there is really no political perspective, and yet this entire project is based on an uncritical acceptance of neoliberal global capitalism and the notion that most forms of government intervention are bad. Since no alternative to capitalism is considered and Big Government is rejected, the only solution is to pursue minor nudges to help individuals make better decisions within the current structure.[10]

As part of its political ideology, behavioral economics often posits that the old "liberal" model of economics does not work because it is based on the misguided notion that people are rational actors who always pursue their best self-interest[11]: "Whether or not they have ever studied economics, many people seem at least implicitly committed to the idea of homo economicus, or economic man—the notion that each of us thinks and chooses unfailingly well, and thus fits within the textbook picture of human beings offered by economists."[12] Behavioral economics thus seeks to counter the way economics is usually taught and practiced by showing how individuals have limited mental capabilities, which prevent them from being fully rational or competent[13]: "If you look at economics textbooks, you will learn that homo economicus can think like Albert Einstein, store as much memory as IBM's Big Blue, and exercise the willpower of Mahatma Gandhi. Really. But the folks that we know are not like that. Real people have trouble with long division if they don't have a calculator, sometimes forget their spouse's birthday, and have a hangover on New Year's Day. They are not homo economicus; they are homo sapiens."[14] Thus, behavioral economists stress the limitations of human cognition, and these limitations call for a more active role for a third party to help out isolated individual decision-makers, but the underlying distrust of large social interventions and government programs pushes these economists to reduce the political to minor changes in how choice opportunities are presented.

While psychoanalysis reveals the ways people are shaped by conflicting drives and unconscious processes, for behavioral economists, the central problem is that people fail to act as rational actors because they are ruled by the automatic parts of their brain: "With respect to diet, smoking, and drinking, people's current choices cannot reasonably be claimed to be the best means of promoting their well-being. Indeed, many smokers, drinkers, and overeaters are willing to pay third parties to help them make better decisions."[15] Since these behavioral economists do not think that our social system can be changed, and they do not want to recommend an expanded role for government or psychoanalysis, the only solution is for people to pay a third party to help influence them in the right direction.[16] Of course, one of the things they do not consider is that people can gain insight into their unconscious and drives and change their self-destructive behaviors through introspection.

Since Thaler and Sunstein recognize that people often do not act with rational self-interest, but the economists also do not appear to believe in psychoanalysis or any other form of therapeutic treatment, they are forced to argue that expert third parties need to nudge people in the right direction in a non-authoritative way: "Unlike Econs, Humans predictably err. Take, for example, the 'planning fallacy'—the systematic tendency toward unrealistic optimism about the time it takes to complete projects. It will come as no surprise to anyone who has ever hired a contractor to learn that everything takes longer than you think, even if you know about the planning fallacy."[17] Here they state that even if people are aware of their own faulty thinking, they will still engage in the same faulty actions, and so the only solution is to find ways to nudge people to do the right thing.

THE SPLIT BRAIN

In order to show why people do not behave as rational self-interested actors, one of the big moves of many behavioral economists is to divide the mind into two discrete systems.[18] On one side, there is the intuitive brain programmed to make fast, automatic decisions, and on the other side, we find the slow reflective brain, which uses conscious analysis:

> Many psychologists and neuroscientists have been converging on a description of the brain's functioning that helps us make sense of these seeming contradictions. The approach involves a distinction between two kinds of thinking, one that is intuitive and automatic, and another that is reflective and rational. We will call the first the Automatic System and the second the Reflective System. (In the psychology literature, these two systems are sometimes referred to as System 1 and System 2, respectively.)[19]

One of the main ways that behavioral economics then borrows from neuroscience is by dividing the mind/brain into two halves, and we shall see how the one part of the brain is equated with the unconscious and the other with conscious reflection.

This splitting of the mind/brain is apparent in how many neuroscientists, evolutionary psychologists, and behavioral economists define the functioning of different brain regions.[20] For instance, Thaler and Sunstein provide the following diagram to map out the difference between the systems:

Automatic system	Reflective system
Uncontrolled	Controlled
Effortless	Effortful
Associative	Deductive
Fast	Slow
Unconscious	Self-aware
Skilled	Rule-following

Nudge, 20

On one level, the major distinction between the two types of mental systems is that one is unconscious and the other is conscious. However, on another level, the distinction appears to be between what is intuitive and automatic and what is the result of individual, reflective cognition. As we saw in the work of Damasio and Pinker, the fast and unconscious system is able to make automatic judgments because it is the result of inherited biological programs, while the slow, deductive system appears to represent a break with nature and evolution. Yet, things are much more complicated than they appear, and this binary opposition between two distinct systems will be proven to be inadequate.

One of the first problems is that when Thaler and Sunstein give examples of automatic thinking, they include thought processes that are neither instinctual nor intuitive. For example, in the following passage, the stress is on the way certain learned knowledge and behavior becomes rote and automatic:

> Most Americans have an Automatic System reaction to a temperature given in Fahrenheit but have to use their Reflective System to process a temperature given in Celsius; for Europeans, the opposite is true. People speak their native languages using their Automatic Systems and tend to struggle to speak another language using their Reflective Systems. Being truly bilingual means that you speak two languages using the Automatic System. Accomplished chess players and professional athletes have pretty fancy intuitions; their Automatic Systems allow them to size up complex situations rapidly and to respond with both amazing accuracy and exceptional speed.[21]

Here we see that the automatic system has to be itself divided in half since part of this system is determined by automatic programs inherited from evolution, while the other half represents learned behavior that has been internalized and memorized and can now function in an automatic way. This double definition of the automatic for neuroscience creates a huge problem because when one does a brain scan and tries to measure the activity in a particular region, one does not know if the lack of energy exerted comes from the fact that the process has been routinized or if it has not been activated.[22]

Like many other behavioral economists and neuroscientists, Thaler and Sunstein both recognize and repress these two very different ways of defining mental automatism. For instance, the following passage appears to collapse the difference between inherited programs and learned routine behaviors:

One way to think about all this is that the Automatic System is your gut reaction and the Reflective System is your conscious thought. Gut feelings can be quite accurate, but we often make mistakes because we rely too much on our Automatic System ... The Automatic System starts out with no idea how to play golf or tennis. Note, however, that countless hours of practice enable an accomplished golfer to avoid reflection and to rely on her Automatic System—so much so that good golfers, like other good athletes, know the hazards of "thinking too much" and might well do better to "trust the gut," or "just do it."[23]

Here the opposition of the two systems breaks down because reflective activities become automatic through practice, and this transfer between the two systems undermines the distinction between nature and nurture. Moreover, the psychoanalytic conception of the unconscious is repressed by equating the unconscious with mental processes that do not require constant conscious attention.

As we saw in the last chapter, evolutionary psychologists rely on distinguishing between what is the result of evolution and what comes from culture and learning, but here we see how these distinctions can evaporate. In fact, a major move of behavioral economics is to explain how humans often make faulty decisions and interpretations because they are prone to simplified rules of thumb or what they call heuristics:

This insight, first developed decades ago by two Israeli psychologists, Amos Tversky and Daniel Kahneman (1974), has changed the way psychologists (and eventually economists) think about thinking. Their original work identified three heuristics, or rules of thumb—anchoring, availability, and representativeness—and the biases that are associated with each. Their research program has come to be known as the "heuristics and biases" approach to the study of human judgment. More recently, psychologists have come to understand that these heuristics and biases emerge from the interplay between the Automatic System and the Reflective System.[24]

In this passage, it appears that the strict division of the mind into two systems is challenged because the common biases detected by behavioral economists are derived from the "interplay" between the two systems. For instance, if I think that homicides are more prevalent than suicides because the media reports on homicides to a greater extent, my automatic brain is using information derived from my reflective mind in a fast and intuitive way.

A further complication of this conception of the divided mind is the way that associative thinking is conceived as something that is both learned and intuitive:

> The third of the original three heuristics bears an unwieldy name: representativeness. Think of it as the similarity heuristic. The idea is that when asked to judge how likely it is that A belongs to category B, people (and especially their Automatic Systems) answer by asking themselves how similar A is to their image or stereotype of B (that is, how "representative" A is of B). Like the other two heuristics we have discussed, this one is used because it often works. We think a 6-foot-8-inch African-American man is more likely to be a professional basketball player than a 5-foot-6-inch Jewish guy because there are lots of tall black basketball players and not many short Jewish ones (at least not these days). Stereotypes are sometimes right![25]

In this description of stereotypes as a mental heuristic, we find a third form of automatic thinking, which is neither based on inherited instincts or routinized behavior but instead points to repeated cultural associations. The problem then is that these three different types of automatic thinking are often confused because they all tend to be fast and non-conscious. Although the authors argue that stereotypes may sometimes be right, the problem is that they are usually at least partially wrong. Moreover, the use of stereotypical thinking represents one of the central aspects of neuroliberalism: people are able to conform to the expected behavior of others because they have reduced other people to knowable stereotypes.[26] Also, we see here how prejudice can take on a scientific quality by being attached to the neuroscientific understanding of the brain and mind.

Since automatic responses, learned behavior, and stereotypes are all tied to the automatic mind, we discover that biology, culture, and psychology are often mixed together to naturalize social constructions from a personal perspective. Here the main goals of neoliberal ideology are accomplished because social hierarchies are not only naturalized, but they are also internalized on a personal basis as repeated stereotypes are given a scientific foundation.[27] We shall see that as Thaler and Sunstein critique the use of facile associations and prejudices, they also argue that they are also inevitable and helpful:

> The picture that emerges is one of busy people trying to cope in a complex world in which they cannot afford to think deeply about every choice they have to make. People adopt sensible rules of thumb that sometimes lead

them astray. Because they are busy and have limited attention, they accept questions as posed rather than trying to determine whether their answers would vary under alternative formulations. The bottom line, from our point of view, is that people are, shall we say, nudge-able. Their choices, even in life's most important decisions, are influenced in ways that would not be anticipated in a standard economic framework.[28]

Due in part to the busy nature of contemporary life, people need to rely on fast rules of thumb to make decisions, but these intuitive heuristics can be misleading, and so there is a need for a third party to nudge people in the right direction.

Although with Pinker's evolutionary psychology and Damasio's neuroscience, we encountered a strong move against social influence in favor of biological determinism, in the case of behavioral economics, the Third Way points to how people internalize social signals on a non-conscious level:

Hundreds of studies confirm that human forecasts are flawed and biased. Human decision making is not so great either. Again to take just one example, consider what is called the "status quo bias," a fancy name for inertia. For a host of reasons, which we shall explore, people have a strong tendency to go along with the status quo or default option. As previous social science research, people not only tend to stick with the status quo, but they also conform to the social group around them even if the group goes against their own beliefs and values.[29]

From this perspective, the unconscious mind is shaped by social factors that are internalized without reflection, and so the unconscious can no longer be seen as a pure derivative of evolution and biology, and yet the political frame of libertarian paternalism pushes for a greatly reduced role for social intervention:

Our hope is that that those recommendations might appeal to both sides of the political divide. Indeed, we believe that the policies suggested by libertarian paternalism can be embraced by Republicans and Democrats alike. A central reason is that many of those policies cost little or nothing; they impose no burden on taxpayers at all.[30]

It should be clear from this passage that this form of economics is being shaped by a political agenda that seeks to find a common ground between liberals and conservatives during a historical period where the belief in government has been greatly reduced. In other terms, their science is clearly political, and even if they consider the moderate middle to be non-ideological, we see how it has internalized the conservative backlash against the welfare state:

> Republicans want to make people's lives better; they are simply skeptical, and legitimately so, about eliminating people's options. For their part, many Democrats are willing to abandon their enthusiasm for aggressive government planning. Sensible Democrats certainly hope that public institutions can improve people's lives. But in many domains, Democrats have come to agree that freedom of choice is a good and even indispensable foundation for public policy. There is a real basis here for crossing partisan divides.[31]

Here we find what Slavoj Zizek has called the technocratic middle, which only offers small reforms instead of challenging the way neoliberal capitalism and politics have produced extreme inequality and poverty.[32]

Therefore, in the face of massive social and ecological problems, behavioral economists tend to focus on making minor adjustments because they do not believe that there is any alternative to our current capitalistic system, and they also do not trust the modern welfare state to distribute resources more fairly or intervene effectively to counter large social problems. The trap then of neoliberalism is that it is a fatalistic discourse, which is impotent in the face of ecological and economic catastrophe. Instead of calling for a stronger state or a fairer economic system, Thaler and Sunstein call for minor psychological manipulations:

> Libertarian paternalism, we think, is a promising foundation for bipartisanship. In many domains, including environmental protection, family law, and school choice, we will be arguing that better governance requires less in the way of government coercion and constraint, and more in the way of freedom to choose. If incentives and nudges replace requirements and bans, government will be both smaller and more modest. So, to be clear: we are not for bigger government, just for better governance.[33]

The problem with this seemingly rational compromise is that it does not allow for any real change to happen, and it fails to confront the enormity of our current problems as it internalizes the conservative backlash against government:

In short, libertarian paternalism is neither left nor right, neither Democratic nor Republican. In many areas, the most thoughtful Democrats are going beyond their enthusiasm for choice-eliminating programs. In many areas, the most thoughtful Republicans are abandoning their knee-jerk opposition to constructive governmental initiatives. For all their differences, we hope that both sides might be willing to converge in support of some gentle nudges.[34]

As a form of micro-economics, this political science does not deal with the effects of global capitalism and the failures of democratic states to provide for their citizens; instead it seeks out a way to influence isolated individuals by finding non-intrusive ways to nudge them in a certain direction.

THE SCIENCE OF SOCIAL CONFORMITY

It is interesting to note that just as Thaler and Sunstein endorse the political and economic status quo, they offer up several different reasons why people tend to conform to their surrounding social environment:

For lots of reasons, people have a more general tendency to stick with their current situation. This phenomenon, which William Samuelson and Richard Zeckhauser (1988) have dubbed the "status quo bias," has been demonstrated in numerous situations. Most teachers know that students tend to sit in the same seats in class, even without a seating chart. But status quo bias can occur even when the stakes are much larger, and it can get us into a lot of trouble.[35]

One reason, then, why people tend to conform to the status quo is that they are afraid of taking risks, and they are averse to loss, and yet, we must ask why do these behavioral economists also conform to the political and economic status quo? In other words, we can read their analysis of social conformity as reflecting on their own arguments and theory since they want to privilege the current backlash against government and the celebration of individual liberty as they acknowledge the limitations of human decision-making.[36]

Behavioral economics is therefore a theory of the limitations of human thinking and planning that is itself a product of its own neuroliberal political ideology. For instance, in the following example, the focus on human inertia feeds into the inertia of their own politics and economics:

The combination of loss aversion with mindless choosing implies that if an option is designated as the "default," it will attract a large market share. Default options thus act as powerful nudges. In many contexts defaults have some extra nudging power because consumers may feel, rightly or wrongly, that default options come with an implicit endorsement from the default setter, be it the employer, government, or TV scheduler.[37]

Just as consumers tend to go with the status quo, we have seen how behavioral economics itself tends to support the neoliberal consensus as it shies away from pointing to the failure of our capitalist system and the need for direct government intervention. Although Thaler and Sunstein desire to turn to neuroscience and evolutionary psychology to present a new scientific and ideology-free version of economics, from a psychoanalytic perspective, we can argue that it is always important to look at how a person's statements are related to their own perspective and desire; one always has to ask why is this person saying this and to whom are they really addressing their discourse.[38] Moreover, how does their perspective on the world shape what they are saying now about the world?

For behavioral economists, a key question concerns the social frame that controls what people see and do not see:

Framing works because people tend to be somewhat mindless, passive decision makers. Their Reflective System does not do the work that would be required to check and see whether reframing the questions would produce a different answer. One reason they don't do this is that they wouldn't know what to make of the contradiction. This implies that frames are powerful nudges, and must be selected with caution.[39]

Since people are seen as lazy, they are easily manipulated by social forces, which frame how people encounter reality. In fact, from a psychoanalytic perspective, individuals never see reality directly, and so there is always a fantasy that frames their perceptions of the world around them.[40] In the case of Thaler and Sunstein, libertarian paternalism is their neuroliberal fantasy frame that distorts their view, and yet this frame is itself one of the objects of their perspective. However, just because they are aware of their own bias, it does not mean that they are not biased because they still conform to their ideological perspective. In fact, Zizek argues that a central aspect of contemporary ideology is that even if people know they are seeing things from a distorted perspective, they still conform to what they see.[41] For instance, since people do not think our economic or political systems can really change, the only thing they can do is to try to outcompete others in a system in which they do not believe. Here we see how cynical conformity is the dominant mode of neoliberal subjectivity: out of fear, people conform, and yet they doubt the value and truth of the social system itself.[42]

In the case of behavioral economists, at the same time they tend to conform to the current market system, they seek to show how consumers are not always making the best choices because they fail to use their reflective minds, and they act impulsively. While the modern economists they critique believed that people were rational, self-interested actors, the contemporary behavioral economists stress the irrational and self-destructive aspects of human nature, and yet their conformity to the present market system is evident: "In many cases, markets provide self-control services, and government is not needed at all. Companies can make a lot of money by strengthening Planners in their battle with Doers, often doing well by doing good."[43] It appears that these behavioral economists trust the market and corporations, but they do not really trust consumers: "Markets provide strong incentives for firms to cater to the demands of consumers, and firms will compete to meet those demands, whether or not those demands represent the wisest choices."[44] Here we see how underlying much of neuroliberal ideology is a distrust of the very individuals that are being set free and celebrated; in other terms, at the same moment when the liberty of the individual is privileged, this individual is shown to be irrational and self-defeating. The result of this contradiction is that we discover that the supposedly free individual is acting on automatic pilot and therefore has to be nudged to do the right thing, but this essential irrationality of humans means that any type of organized social planning will be destructive and ineffective because it is guided by irrational individuals.[45]

Since these behavioral economists do not think that irrational people can come up with rational solutions to large social problems, they offer a cost-free, bipartisan political agenda. Once again, we enounter here how the conservative backlash against the postmodern welfare state is seen as the natural, default ground for political discourse; since everyone now has internalized the Republican message that taxes are a form of theft and government is the problem and not the solution, the government must find ways to influence people without spending money. *Nudge* therefore helps us to understand how different academic disciplines can help rationalize and naturalize the political and status quo, and while these discourses pretend to be unbiased and scientific, they reveal deep political commitments. Although the libertarian paternalistic perspective would have been considered to be extremely conservative 50 years ago, these behavioral economists are now able to claim that they represent the modern middle ground, which is itself represented as being void of ideology and bias.

THE LIBERAL IN THE NEUROLIBERALISM

I have been arguing that one reason why we can consider behavioral economics to be a neoliberal discourse is that it participates in the conservative rejection of "Big Government" and the welfare state as it focuses on how to nudge people who fall prone to faulty thinking and self-destructive behavior. However, there is also a liberal aspect of neoliberalism, which we find in the way that individuals seek to conform to the social order in order to have their ideal selves recognized by an ideal Other. In this cultural and psychological structure, liberals invest in a meritocratic system that judges people based on their talent and education and not on their inherited social position.[46] One way then that liberals participate in neuroliberalism is by using competition to validate individual worth. As we shall see in the case of behavioral economics, a key aspect of this liberal ideology is the way individuals conform to an idealized social order to have their own ideal self recognized by an ideal Other. For Lacan, this structure represents the essence of secondary narcissism, the idealizing transference, and the fundamental structure of the Imaginary: the ideal ego seeks to be verified by an idealized Other.[47]

LACAN'S CRITIQUE OF LIBERAL CONFORMITY

Lacan's critique of narcissism and what he called the American way of life is in part derived from his argument that the European analysts who fled to America from Europe after World War II sought to assimilate to the US culture, and this effort at social conformity resulted in a psychoanalytic practice of ritualized identification.[48] Lacan insisted that the American brand of ego psychology saw the goal of analysis as based on the identification of the patient's ego with the ideal ego of the analyst. In other words, the American analysts turned psychoanalysis into a theory and practice of social conformity, which itself acted to repress the unconscious and the Freudian notion of sexuality.

One way that the American ego psychologists moved away from Freud was through their idea that the goal of analysis is to strengthen the healthy part of the ego and to avoid conflict. For Lacan, this focus on assimilation and adaptation fit in well with the American emphasis on leading a happy life by fitting into the rest of society. Lacan also argued that the ahistorical nature of American culture helped to turn American psychoanalysis and therapy away from Freud's most profound insights into memory and trauma. Lacan attached this repression of analysis by analysts to the dominance of behaviorism and the focus on conformity and conditioning.

In the work of Thaler and Sunstein, we can find a similar influence of behaviorism on their conceptions of social conformity. For instance, in their description of different research experiments and findings, they point to a mode of adaptation that eliminates a consideration of individual subjectivity:

> For a quick glance at the power of social nudges, consider just a few research findings: 1. Teenage girls who see that other teenagers are having children are more likely to become pregnant themselves. 2. Obesity is contagious. If your best friends get fat, your risk of gaining weight goes up. 3. Broadcasters mimic one another, producing otherwise inexplicable fads in programming. (Think reality television, American Idol and its siblings, game shows that come and go, the rise and fall and rise of science fiction, and so forth.) 4. The academic effort of college students is influenced by their peers, so much so that the random assignments of first-year students to dormitories or roommates can have big consequences for their grades and hence on their future prospects. (Maybe parents should worry less about which college their kids go to and more about which roommate they get.) 5. Federal judges on three-judge panels are affected by the votes of their colleagues. The typical Republican appointee shows pretty liberal voting patterns when sitting with two Democratic appointees, and the typical Democratic appointee shows pretty conservative voting patterns when sitting with two Republican appointees. Both sets of appointees show far more moderate voting patterns when they are sitting with at least one judge appointed by a president of the opposing political party. The bottom line is that Humans are easily nudged by other Humans. Why? One reason is that we like to conform.[49]

In all aspects of life, we find individuals conforming to other people around them, and we can read this as proof of the effectiveness of nudging people in a certain direction and the lack of self-awareness and self-control of the isolated individual. To prove this power of social conformity, as shown by the research of Solomon Asch, these behavioral economists turn to neuroscience: "Remarkably, recent brain-imaging work has suggested that when people conform in Asch-like settings, they actually see the situation as everyone else does."[50] In this use of neuroscience, we see how just relating a finding to a brain imaging study gives it a greater sense of credibility, and in this case, the social conformity of individuals results in a conformity to their neural activity. Moreover, Thaler and Sunstein refer here to the disputed neuroscience discovery of mirror neurons, which represent a way of displacing the older psychoanalytic notions of narcissism and identification.[51]

It is important to stress that unlike neuroscientists and evolutionary psychologists, who tend to discount the autonomy of the social order and social conformity, behavioral economists often stress social influences, but they divorce the social from individual psychology or the psychoanalytic version of unconscious processes. For instance, in the following description of the conservative tendencies of collective behaviors, the social is seen as an autonomous realm:

> We can see here why many groups fall prey to what is known as "collective conservatism": the tendency of groups to stick to established patterns even as new needs arise. Once a practice (like wearing ties) has become established, it is likely to be perpetuated, even if there is no particular basis for it. Sometimes a tradition can last for a long time, and receive support or at least acquiescence from large numbers of people, even though it was originally the product of a small nudge from a few people or perhaps even one. Of course, a group will shift if it can be shown that the practice is causing serious problems. But if there is uncertainty on that question, people might well continue doing what they have always done.[52]

Although it is common to see social conformity as inherently conservative, I have been arguing that neoliberalism reveals a liberal aspect of group thinking: during a time of rapid social and economic change that is being led by a conservative push for free markets and reduced government, it is often the liberals who seek to maintain social order through social conformity to older models of collective organization (unions, tenure, welfare).

In a meritocratic culture, liberal elites want the status quo to be maintained because they see themselves as the benefactors of a system that privileges education and recognizes talent.[53] Moreover, this type of social order emphasizes a narcissistic desire to have the ideal ego recognized by an ideal Other.[54] However, for Thaler and Sunstein, social narcissism is removed from its connection to the psychoanalytic understanding of Imaginary rivalry, duality, and jealousy: "One reason why people expend so much effort conforming to social norms and fashions is that they think that others are closely paying attention to what they are doing. If you wear a suit to a social event where everyone else has gone casual, you feel like everyone is looking at you funny and wondering why you are such a geek." As psychoanalysis explains, narcissism means that people are obsessed about what others think about them, and so they need to have their ideal self verified by an external Other; yet for Thaler and Sunstein, the causes for individual behavior are derived from automatic responses derived from biology or social conformity and not unconscious thinking.

One way, then, the neuroliberalism represses the unconscious is by focusing on how people copy social behaviors on the level of cynical conformity, which allows for the elimination of both conscious free will and repressed unconscious formations: "An important problem here is 'pluralistic ignorance'—that is, ignorance, on the part of all or most, about what other people think. We may follow a practice or a tradition not because we like it, or even think it defensible, but merely because we think that most other people like it."[55] Unlike conservatives who believe in what they believe, the idea here is that cynical conformists do not have to believe or even understand their own beliefs and social actions. For Zizek, this type of conformity and belief is the key to neoliberal subjectivity; in a mass society, people follow the social beliefs of others from a distance, and what makes someone a fundamentalist is that they really believe in their beliefs and that is why they pose such a threat.[56] According to Zizek, what the liberal postmodern multiculturalists fear is any strong commitment to a particular identity or belief.

It is important to stress that like behaviorism, neoliberal cynical conformity functions by establishing a connection between the individual and the social by placing individual thinking in a black box: since humans are seen as being conditioned by social cues without concern for thoughts and feelings, the behaviorist is able to make a direct connection between the social and the self.[57] Here we witness how behavioral economics is able to remove the concept of repression from the unconscious: since repression entails an individual withholding information from himself or herself, it is centered on an internal dynamic, while behaviorism excludes the internal from consideration. Moreover, cynical conformity represents a central method for repressing repression because it is centered on the idea that one can escape one's own unconscious by merely copying the actions of others without any internal processing.[58]

For behavioral economists like Thaler and Sunstein, cynical conformity also helps us to understand how economic bubbles tend to be created:

> The best account has been given by Robert Shiller, who emphasizes the role of psychological factors and herd behavior in volatile markets. Shiller contends that "the most important single element to be reckoned with in understanding this or any other speculative boom is the social contagion of boom thinking, mediated by the common observation of rapidly rising prices." He urges that in the process of social contagion, public knowledge is subject to a kind of escalation or spiral, in which most people come to think that optimistic view is correct simply because everyone else seems to accept it.[59]

While behavioral economists see social influence as taking advantage of automatic mental processes, Freud argues that social contagions are derived through the unconscious process of identification and symptom formation.[60] Moreover, Freud insists that the material of unconscious processes are thoughts that were previously conscious; however, for behavioral economists, neuroscientists, and evolutionary psychologist, consciousness is often neglected or relegated to an effect and not a cause of behavior.

When behavioral economists do turn to conscious thought, it is usually to focus on how it fails to control faulty automatic thinking:

> There is a general point here. If consumers have a less than fully rational belief, firms often have more incentive to cater to that belief than to eradicate it. When many people were still afraid of flying, it was common to see airline flight insurance sold at airports at exorbitant prices. There were no booths in airports selling people advice not to buy such insurance.[61]

Instead of following the usual libertarian endorsement of the free market, we see here how the mistrust of the government can be coupled with a mistrust of markets, and the result is a call for technocratic experts to step in and try to nudge people in the right direction. We also encounter here how the conservative criticism of Big Government can be combined with a liberal criticism of the economy; in this case, the neoliberal consensus is based on a distrust of all social systems, and yet the focus is on the power of social influence.

The desire to steer a middle ground between conservatives and liberals push Thaler and Sunstein to both accept and reject both sides of the political spectrum. For instance, in discussing the role of subprime loans in the 2008 financial crisis, they make the following seemingly balanced argument:

> As is often the case, there are two extreme views about subprime loans. Some, particularly those left of center or in the news media, label all such loans with the derogatory term predatory. This broad brush fails to recognize the obvious fact that higher-risk loans will have to have higher interest rates to compensate the people who lend the money. The fact that poor and risky borrowers pay higher interest rates does not make these loans "predatory." In fact, the microfinance loans in developing countries that led to a well-deserved Nobel Peace Prize for Muhammad Yunus in 2006 often come with interest rates of 200 percent or more, yet the borrowers are made better off by these loans. On the other side, some observers think that the hue and cry about predatory lending is based entirely on the failure of left-leaning journalists and others to understand that risky loans require higher interest rates. As usual, the truth lies somewhere between the two extremes. Subprime lending is neither all good nor all bad.[62]

This balanced approach prevents them from offering any strong critique of our financialized capitalist system, and it also blinds them from seeing how far right the middle has shifted. In the case of subprime loans, it is simply accepted that low-income people should have to pay a price for their poverty, and what they never ask is why do so many people have to rely on this type of financing. In fact, when they do examine the increased level of debt for individual Americans, they fail to point out the larger structural features of the American economy: "On average, American households spent more than they earned and borrowed more than they saved. Increased borrowing rates were fueled by substantial growth in home equity loans and in credit card debt."[63] What is missing from this analysis is the roles played by stagnant wages, wealth inequality, racism, globalization, and outsourcing. Like other neuroliberal discourses, the focus tends to be on the private individual who makes bad decisions because of defective thinking and behavior.

Instead of examining the larger economic trends that feed economic desperation, these reformists seek to offer small nudges to deal with macro-economic factors:

> Some demand an end to predatory lending, but because loans do not come stamped "predatory," it is hard to implement any such ban without depriving many deserving but high-risk borrowers from any source of financing. And of course, we libertarian paternalists do not favor bans. Instead, we prefer an improvement in choice architecture that will help people make better choices and avoid loans that really are predatory—loans that exploit people's ignorance, confusion, and vulnerability.[64]

On one level, it looks like the authors are being supportive of low-income people who are forced to engage in risky, high-interest loans, but on another level, we see that these vulnerable people are being blamed for their ignorance and confusion. Like so many other aspects of neoliberal ideology, the focus here is on the individual who is the center of blame and responsibility, and yet we have also seen that individual thinking and consciousness is often excluded from consideration as the emphasis is on automatic processes and social conformity.

This focus on the individual consumer and the failure to take into consideration the economic and political results of neoliberal policies is revealed in the following discussion of student debt:

> The cost of going to college has been rising almost as fast as the cost of health care and rare baseball cards. At many private universities, including ours, it costs a student more than fifty thousand dollars a year in tuition, room, and board. Scholarships and part-time jobs typically do not cover the cost of college. So students and their families often turn to student loans to help out. In fact, loans are a common option. About two-thirds of four-year college students are in debt when they graduate.[65]

What these behavioral economists do not ask is why have colleges and universities raised their tuition so much, and why we do not have a better-funded system for higher education.[66] Since they tend to accept the political and economic status quo, the only thing they can do is to provide small, technocratic solutions to large social problems, and once again, part of this status quo is the idea that individuals and not the government should pay for things like higher education.

At the same time that Thaler and Sunstein want to be seen as bipartisan by offering a third way between conservatives and liberals, they reveal how the middle has shifted so far to the right that they see no problem taking a balanced approach to privatizing social security, creating more charter schools, and reducing the right to sue for medical malpractice. All of these conservative backlash issues have been internalized as the status quo, and so their neutral frame is really one profoundly shaped by contemporary neoliberal politics. For example, in their discussion of the New Deal, we see how they have accepted the conservative framing of social welfare policies: "Ever since Franklin Delano Roosevelt's New Deal, the Democratic Party has shown a great deal of enthusiasm for rigid national requirements and for command-and-control regulation. Having identified serious problems in the private market, Democrats have often insisted on firm mandates, typically eliminating or at least reducing freedom of choice."[67] By equating programs like social security with "command-and-control regulation," they are clearly repeating the conservative backlash rhetoric that equates any government program with tyranny and totalitarianism.[68] Moreover, in their discussion of the Republican reaction to the welfare state, they fail to articulate many of the underlying goals of the conservative counter-revolution: "Republicans have responded that such mandates are often uninformed or counterproductive—and that in light of the sheer diversity of Americans, one size cannot possibly fit all. Much of the time, they have argued on behalf of laissez-faire and against government intervention. At least with respect to the economy, freedom of choice has been their defining principle."[69] What is missing from this account is the desire of wealthy people to not pay taxes and their effort to undermine the ability of the government to regulate their businesses.[70]

Although it may seem that I have moved away from the role played by neuroscience and evolutionary psychology in shaping neoliberal ideology, what we find in behavioral economics is a shared acceptance of the current political and economic system coupled with a focus on how individuals are guided by non-conscious forces. Whether the emphasis is on genes or social memes, we still end up with a reinforcement of the status quo as any real alternative is dismissed for scientific reasons. In fact, as we will see in the next chapter, neuroscience, evolutionary psychology, and behavioral economics are often combined together to produce neuroliberalism.

NOTES

1. Thaler, Richard, and Cass Sunstein. "Nudge: The gentle power of choice architecture." *New Haven, Conn.: Yale* (2008).
2. Narveson, Jan. *The libertarian idea.* Broadview Press, 2001.
3. Frank, Thomas. *One market under God: Extreme capitalism, market populism, and the end of economic democracy.* Anchor Canada, 2001.
4. Hayek, Friedrich August, and Bruce Caldwell. *The road to serfdom: Text and documents: The definitive edition.* Routledge, 2014.
5. *Nudge*, 5.
6. Ibid.
7. Ibid.
8. Giddens, Anthony. *The third way and its critics.* John Wiley & Sons, 2013.
9. Zizek, Slavoj. "Multiculturalism, or, the cultural logic of multinational capitalism." *New left review* 225 (1997): 28.
10. Fisher, Mark. *Capitalist realism: Is there no alternative?* John Hunt Publishing, 2009.
11. Ariely, Dan. *Predictably irrational.* New York: HarperCollins, 2008.
12. *Nudge*, 6.
13. Kahneman, Daniel. *Thinking, fast and slow.* Macmillan, 2011.
14. *Nudge*, 7.
15. Ibid.
16. Heukelom, Floris. *Behavioral economics: a history.* Cambridge University Press, 2014.
17. *Nudge*, 7.
18. Kahneman.
19. *Nudge*, 19.
20. McGilchrist, Iain. *The master and his emissary: The divided brain and the making of the western world.* Yale University Press, 2009.
21. Ibid.
22. Friston, Karl J. "Modes or models: a critique on independent component analysis for fMRI." *Trends in cognitive sciences* 2.10 (1998): 373–375.
23. *Nudge*, 21.
24. Ibid., 23.
25. Ibid., 26.
26. Brooks, David. *The social animal: The hidden sources of love, character, and achievement.* Random House Incorporated, 2012.
27. Duggan, Lisa. *The twilight of equality?: Neoliberalism, cultural politics, and the attack on democracy.* Beacon Press, 2012.
28. *Nudge*, 37.
29. Ibid., 7.
30. Ibid., 13.

31. Ibid., 14.
32. Žižek, Slavoj. *In defense of lost causes*. Verso, 2009.
33. *Nudge*, 14.
34. Ibid.
35. Ibid., 34.
36. Camerer, Colin, et al. "Regulation for Conservatives: Behavioral Economics and the Case for "Asymmetric Paternalism"." *University of Pennsylvania Law Review* 151.3 (2003): 1211–1254.
37. *Nudge*, 35.
38. Lacan, J. "Direction and Power of Treatment." *Écrits: A Selection* (2001): 250–310.
39. *Nudge*, 37.
40. Žižek, Slavoj. *The plague of fantasies*. Verso, 1997.
41. Žižek, Slavoj. *The sublime object of ideology*. Verso, 1989.
42. Bloom, Peter. "Capitalism's Cynical Leviathan: Cynicism, Totalitarianism, and Hobbes in Modern Capitalist Regulation." *International Journal of Žižek Studies* 2.1 (2008).
43. *Nudge*, 49.
44. Ibid.
45. Hayek, Friedrich August. "The use of knowledge in society." *The American Economic Review* 35.4 (1945): 519–530.
46. Frank, Thomas. *Listen, Liberal: Or, What Ever Happened to the Party of the People?*. Macmillan, 2016.
47. Lacan, Jacques. "Remarks on Daniel Lagache's presentation: 'psychoanalysis and personality structure'." *Écrits [1961]* (2006): 543–574.
48. In the English version of Lacan's *Écrits*, one can trace the development of his critique of American culture, social conformity, and ego psychology (38, 115, 127–28, 231, 243).
49. *Nudge*, 55.
50. Ibid., 57.
51. Hickok, Gregory. *The myth of mirror neurons: The real neuroscience of communication and cognition*. WW Norton & Company, 2014.
52. *Nudge*, 58.
53. Hayes, Christopher. *Twilight of the elites: America after meritocracy*. Broadway Books, 2013.
54. Verhaeghe, Paul. *What about Me?: the struggle for identity in a market-based society*. Scribe Publications, 2014.
55. *Nudge*, 59.
56. Zizek, Slavoj. "Multiculturalism, or, the cultural logic of multinational capitalism." *New left review* 225 (1997): 28.
57. Zizek, Slavoj. "Multiculturalism, or, the cultural logic of multinational capitalism." *New left review* 225 (1997): 28.

58. Zizek, Slavoj. "The interpassive subject." *Traverses [web page] Retrieved December 7* (1998): 2011.
59. *Nudge*, 60.
60. Freud, Sigmund. *Group psychology and the analysis of the ego*. No. 770. WW Norton & Company, 1975.
61. Ibid., 80.
62. *Nudge*, 135.
63. Ibid.
64. Ibid., 137.
65. Ibid., 138.
66. Samuels, Robert. "Why public higher education should be free." (2013).
67. *Nudge*, 252.
68. Hayek, Friedrich August, and Bruce Caldwell. *The road to serfdom: Text and documents: The definitive edition*. Routledge, 2014.
69. *Nudge*, 253.
70. Frank, Thomas. *What's the matter with Kansas?: how conservatives won the heart of America*. Macmillan, 2007.

CHAPTER 5

The Brain Sciences Against the Welfare State

Abstract In order to explore how neoliberal ideology shapes theories and practices of the new brain sciences, this chapter closely reads a study that integrates neuroscience, evolutionary psychology, and behavioral economics. One of the goals of this analysis is to understand the basic underlying theory that connects the different neuroliberal discourses, and although many individual scientists might reject my characterization of their fields, I will argue that their work is only possible because of the belief in the idea of discoverable biological determinism. In short, neuroscientists, evolutionary psychologists, and behavioral economists must accept the rejection of the psychoanalytic unconscious in order to promote a theory of non-conscious thinking that is derived from mental programs determined by evolutionary forces. Moreover, this mode of new social Darwinism is reliant on the idea that inherited mental traits can be connected to specific brain regions, neurons, and genes, and even when there is a recognition of the roles played by culture, learning, experience, and subjectivity, the need to articulate a universal form of human nature overrides all other sources that could shape thinking and behavior.

Keywords Evolutionary psychology • Neuroscience • Neoliberalism • Behavioral economics • Psychoanalysis

In order to explore how neoliberal ideology shapes the theories and practices of the new brain sciences, I will read closely a study that integrates

© The Author(s) 2017
R. Samuels, *Psychoanalyzing the Politics of the New Brain Sciences*,
https://doi.org/10.1007/978-3-319-71891-0_5

neuroscience, evolutionary psychology, and behavioral economics. One of the goals of this analysis is to understand the basic underlying theory that connects the different neuroliberal discourses, and although many individual scientists might reject my characterization of their fields, I will argue that their work is only possible because of the belief in the idea of discoverable biological determinism. In short, neuroscientists, evolutionary psychologists, and behavioral economists must accept the rejection of the psychoanalytic unconscious in order to promote a theory of non-conscious thinking that is derived from mental programs determined by evolutionary forces. Moreover, this mode of new social Darwinism is reliant on the idea that inherited mental traits can be connected to specific brain regions, neurons, and genes, and even when there is a recognition of the roles played by culture, learning, experience, and subjectivity, the need to articulate a universal form of human nature overrides all other sources that could shape thinking and behavior.

THE QUESTION OF WELFARE

In their research paper, "Who Deserves Help? Evolutionary Psychology, Social Emotions, and Public Opinion about Welfare," Michael Bang Petersen, Daniel Sznycer, Leda Cosmides, and John Tooby seek to study how current attitudes about social welfare programs are shaped by inherited mental programs derived from our hunter-gatherer ancestors. In fact, the main focus of this study is to ask from a scientific perspective why people reject or support social welfare policies. To construct this research, we shall see that a whole series of assumptions have to be made, and these assumptions are shaped by an unacknowledged neoliberal political ideology. Moreover, I will argue that this work is a great example of how the new brain sciences help to naturalize the social and economic status quo as they hide political beliefs behind scientific explanations.

Although scientists want to maintain the air of being unbiased, neutral, objective, and universal, what we often find is that their perspectives and research are determined by how they define key terms.[1] For instance, at the start of this study, we are told the following: "When individuals form opinions about social welfare, a primary concern is whether welfare recipients *deserve* the benefits they receive (Cook & Barrett, 1992; Gilens, 1999; Iyengar, 1991; Larsen, 2006; Petersen, Slothuus, Stubager, & Togeby, 2011; Sniderman, Brody, & Tetlock, 1991, chap. 5)."[2] By listing so many other researchers here, it appears that it is common and even natural to argue that popular opinions about welfare programs are shaped by how deserving someone thinks a recipient might be. However, it can also be shown that mainly conservatives focus on this factor, while liberals tend to focus on issues like justice or fairness.[3] In other words, at the very start of their research, these scientists make a guiding assumption that is itself an ideological judgment. Instead of thinking of social welfare programs in terms of the greater good, the focus is on how individuals apply moral judgments concerning the amount of help other people deserve.

The next step in the study is to determine how people decide if someone is deserving of help, and here we find another leap and assumption:

> In deciding whether recipients deserve welfare, individuals pay attention principally to the recipients' efforts in alleviating their own need (Gilens, 1999; Oorschot, 2000). If welfare recipients are seen as able to work, but preferring not to (i.e., they are "lazy"), they are perceived as undeserving and welfare is opposed. In contrast, if welfare recipients are seen as unlucky victims of external circumstances, they are perceived as deserving and welfare is supported.[4]

The important move here is to posit that the main way people decide on the issue of whether other people should receive welfare benefits is by forming a judgment of the laziness of the receiver. Of course, this hypothesis just happens to match the neoliberal conservative rhetoric of the "Welfare Queen" and the idea that government policies like unemployment insurance make people not want to work.[5] Furthermore, it is taken for granted that public policy should be based on the moral judgments of individuals. In fact we should ask, why did these researchers decide to study this particular issue, and how does their research hypothesis reveal an underlying political belief?

In referring in part to the work of one of the authors own previous research, the study argues that this focus on the morality of helping others can be given an empirical scientific foundation: "While strong evidence has been produced that establishes an empirical link between welfare opinions and judgments of recipients' effort, extant research lacks empirically well-supported explanations for why and how these judgments so strongly color welfare opinions."[6] As we saw in the previous chapters, the selection of topics is never purely innocent in the sense that it takes a certain mindset to decide to study the relation between welfare programs and individual moral judgments.

Universal Human Nature

A common feature that this particular research shares with neuroscience and evolutionary psychology is the tendency to posit universal mental processes that go beyond particular cultures; after all, to prove that a reaction to something is caused by biological programs selected by evolutionary forces, it is necessary to posit universal aspects of human nature. In this particular case, the argument is that in all cultures, the researchers find the same connection between opinions about welfare programs and determinations of the laziness of the recipients: "Analyses of cross-cultural data from the World Values Survey show that the perception that poverty is caused by laziness—i.e., a lack of motivation to put in effort—is a universal driver of opposition to government efforts to reduce poverty (see Appendix)."[7] This claim of a universal attitude is full of unspoken assumptions and feeds into the neoliberal conservative backlash against the welfare state. Furthermore, the argument is that since there is a natural and universal desire not to give people aid because they may be lazy, then it is inevitable that welfare state provisions will be reduced. Some important related questions that are not examined are, how do people know if the receivers of aid are lazy and what constitutes poverty and how are different welfare programs evaluated?[8] Of course, none of these questions are posed by this research because the focus is on universal human nature and not on examining specific social systems.

Psychoanalysis Versus Biological Determinism

As I have argued throughout this book, the emphasis on universal human nature conflicts with the psychoanalytic idea that our biology and natural instincts are disrupted by culture and subjectivity.[9] Against the claim of universality, psychoanalysis posits that human sexuality represents both an excess and a lack: since anything can be sexualized, sex itself has no natural essence.[10] Likewise, the unconscious reveals the ways individuals go against natural and cultural determinations: by refusing to distinguish between fact and fantasy and pleasure and pain, the unconscious represents a limit to universal human nature. Moreover, the Imaginary foundation of conscious points to a level of individual mental autonomy that undermines the quest for universal human nature.[11] However, in order to posit a form of psychology unaffected by culture, individuality, and the unconscious, the authors of this study return to a pre-psychoanalytic understanding of human nature.

Although they acknowledge cultural and ideological differences, evolutionary psychologists tend to focus on the underlying mental mechanisms that are determined to be biological and universal. For instance, in looking at data from 49 different countries, the researchers found that public support for welfare programs is tied to the citizens' perceptions of the laziness of the receivers of aid.[12] Once again, the goal here is to make a universal claim so that cultural differences do not play a role, and since culture can be excluded, there must be a biological explanation for the shared human nature.[13]

Against Ideology

One possible cultural cause for popular attitudes toward welfare programs is political ideology, yet the authors quickly dismiss this role for culture and social influence: "several observations suggest that the preoccupation with the effort of needy individuals is grounded in psychological processes that preexist ideology."[14] The aim of this argument is to dismiss the power of ideology so that a universal biological program can be discovered, and yet it is hard to imagine how political and social beliefs can be completely excluded from the analysis. It is also debatable if political attitudes can be seen as purely psychological processes occurring before ideology; in fact, it is questionable whether we can locate a pre-ideological vacuum for political interpretations of governmental programs.[15] Furthermore, this study itself may be a product of a particular historical ideology that attempts to deny its own ideological import.[16] In this sense, ideology here represents a form of repression where the mediating discourse is itself denied.

In order to prove that ideology is not the driving cause behind people's opinions concerning welfare programs, the researchers claim that people of different ideological backgrounds share the same basic attitudes toward the question of who deserves social aid:

First, if people engage in deservingness judgments from some culturally specific ideology, then people who have different or opposing ideologies ought not to provide parallel judgments. Yet they do. In a recent study, Petersen et al. (2011) demonstrated that while egalitarians and nonegalitarians might disagree in the abstract about welfare recipient deservingness, ideological differences vanish when asked to judge the deservingness of specific welfare recipients. In achieving this effect, deservingness judgments were shown to operate in an automated fashion, picking up cues and informing welfare opinions effortlessly (Petersen et al., 2011).[17]

The researchers make several key moves here: first they argue that people with opposing political viewpoints tend to make the same kind of judgments, then they add that these judgments are fast and therefore intuitive. One reason why they need to stress the universal and automatic aspects of these popular reactions to welfare policies is that the theory of evolutionary psychology requires a mental process to be automatic and universal in order to be considered biological and derived from evolution. Furthermore, as we saw in Pinker's work, there is also a need to undermine any cultural or learning-based explanation that could compete with biological determinism.

A goal, then, of this work is to show how social judgments are in actuality responses preprogrammed by evolution:

> We propose that an approach that draws on recent findings in evolutionary psychology and hunter-gatherer studies can help explain why people spontaneously connect support for welfare to welfare recipients' effort. Accumulating evidence from evolutionary psychology and neuroscience indicates that human nature—our universal, reliably developing psychological architecture—includes an array of evolved cognitive and emotion programs tailored by natural selection to solve recurrent adaptive problems faced by our group-living ancestors.[18]

In the articulation of the foundations of evolutionary psychology, we see that the claim for universal human nature is tied to an interpretation of our ancestors' hunter-gatherer culture. As many critics have pointed out, a major problem occurs here because the evolutionary psychologists have to make a lot of assumptions about a form of society that occurred in the unrecorded distant past.[19] In fact, some have suggested that these "scientists" simply recreate a past culture in order to justify their current interpretation of human psychology.[20] No matter what, it should be clear that they are using a version of the past in order to explain present feelings and attitudes, and from this perspective, we encounter the inherent conservatism of this discourse: since they employ an imagined past to define the current status quo, they posit that humans are inherently conservative and traditional. Of course, Freud also argued that people rewrite the past from the perspective of their present knowledge, but he believed that one of the goals of analysis is to distinguish among the different interpretations of past experiences.

In this particular study, the imagined past is constructed in order to show how all people have inherited a set of social beliefs derived from the hunter-gatherer culture:

Here we explore the implications of (1) the hypothesis that the social emotions of anger and compassion were designed by natural selection, in part, to regulate whether and to what extent we want to help a needy person, and (2) the hypothesis that these emotion programs are embedded in a system of cognitive mechanisms that collectively implement a logic of social exchange that evolved to advantageously manage mutual assistance among our ancestors in small-scale foraging groups. From this theoretical framework, we argue that the pervasive effect of perceptions of welfare recipients' effort regarding work on support for welfare arises because these perceptions fit the input systems (i.e., resemble the triggers) that the two social emotion programs, anger and compassion, are designed to monitor and respond to.[21]

Underlying this argument is the idea that it takes a very long time for universal human nature to be selected by evolution, and so our inherited mental programs must have been derived from the long period we spent in hunter-gatherer culture.[22] Moreover, they assume that our basic social responses and interpretations have not been affected by thousands of years of cultural transformation.

Forgetting Culture

In order to prove that current welfare opinions are determined by inherited mental programs, the researchers perform the following studies:

In Study 1, we demonstrate that welfare opinions are in fact powerfully shaped by social emotions. When activated, these emotion programs influence not only welfare opinions but also the ease with which they are formed. In Study 2, we provide evidence that anger and compassion mediate the link between effort cues and attitudes about welfare and that they do so independently of the ideology of the observers. In Study 3 we show that the fit between anger and compassion on the one hand and perceptions of welfare recipients' effort on the other are highly specific. That is: (1) it is specifically perceptions about welfare recipients' effort rather than other types of perceptions that regulate anger and compassion; and (2) perceptions of welfare recipients' effort regulate the activation of anger and compassion rather than emotions such as anxiety, contempt, and disgust. Study 4 demonstrates that this fit between effort perceptions and anger and compassion is robust across two highly different countries and welfare systems: the United States and Denmark.[23]

Before I examine in detail these different studies, it is important to stress that their focus on anger and compassion greatly limits the possible responses individuals can have to welfare programs. Once again, we should examine the particular definitions and variables that shape the research and which can be traced to specific unacknowledged ideological assumptions.

On one level, it is clear that the researchers want to argue for a universal set of psychological responses determined by evolution, but they also signal that they are not excluding social and cultural factors:

> Such an analysis presupposes rather than undermines the importance of other kinds of analyses. In understanding whether a specific individual perceives welfare recipients as lazy or not, one needs to rely on analyses of, for example, the content of the individuals' ideology (Skitka & Tetlock, 1993), the agenda of the media (Gilens, 1999), the structure of political institutions (Larsen, 2006), and the level of ethnic diversity in the individual's country (Alesina et al., 2001). Our contribution here is to illuminate the psychological mechanisms that process such individual and contextual factors and cause them to influence welfare opinions.[24]

Although they have previously discounted the importance of ideology, they now claim that ideology plays a role in shaping the opinions of individuals; the reason for this apparent contradiction is that they want to argue that the basic response mechanisms are universal, but the particular responses are shaped by culture. In other words, the idea is that we all respond with either anger or compassion to social welfare programs, but our use of these emotions is triggered by different ideological perspectives.

The underlying contradiction that runs throughout this study and similar research in evolutionary psychology is the relation between nature and nurture. If all they are saying is that our basic emotional responses are derived from evolution, but the way we respond to particular situations is shaped by experience, learning, and culture, then there would be no need to discount the other social sciences and cultural theorists. The idea here would be similar to the notion that sexual organs are determined by evolution, but the interpretation of gender roles comes from culture.[25] Yet, as we saw in Pinker's work, this dialectic between nature and nurture is not fully endorsed by evolutionary psychology because they also want to discredit the power of current culture and political interventions to shape universal human nature.

To begin their process of distinguishing evolutionary psychology from the other social sciences, they tie emotions to natural selection:

> our emotions constitute an array of distinct and sophisticated information-processing mechanisms, each designed by natural selection to solve specific problems facing our ancestors (Petersen, 2010; Sell et al., Tooby et al., 2008) millions of years, our ancestors lived in small-scale groups (Alford & Hibbing, 2004; Cosmides & Tooby, 2006; de Waal, 1989), and some emotions—the social emotions—have design features that evolved for successfully solving recurrent adaptive problems of group living—such as sharing, exploitation, coalitions, power relations, hierarchy, collective action, punishing norm violators and managing intergroup relations (Petersen, 2009).[26]

An essential step made here is to equate emotions with information-processing programs determined by natural selection to solve particular problems that our ancestors dealt with millions of years ago. In using the phrase "information-processing," we return to the idea that the brain is a computer preprogrammed by evolution; moreover, the assumption is that emotions were first developed to help people find the best way to adapt to a particular environment (hunter-gatherer culture). In tying evolution to computers and our earliest form of social organization, the present is tied to the past, and a continuity is determined between the brain and the mind and between the past and the present.

The Psychoanalysis of Emotions

Like many neuroscientists, the effort by evolutionary psychologists to connect brains, minds, and evolution requires narrowly defining particular mental processes; however, a psychoanalytic understanding of emotions offers a very different view of how emotions actually work. From a psychoanalytic perspective, the human mind is structured by a web of personal and social associations that have been retained through memories.[27] For example, if I have a dream that produces anxiety, we can trace this emotional response to a series of past experiences and interpretations of those experiences. In fact, Freud's *Interpretation of Dreams* reveals how each of the dream elements explored can be linked to past thoughts, and these thoughts are derived from how a particular individual has interpreted a specific situation on an affective, cognitive, and ethical level.[28] For instance, when Freud dreams about his encounter with one of his former patients,

Irma, he shows that his anxiety in the dream revolves around a series of encounters with other patients, and the unifying factor is that in each case, the patient resisted Freud's proposed solution.[29] Here we see how the unconscious automatically makes connections among diverse events that all share a similar set of personal responses and interpretations. It is vital to stress that these emotional responses contain ethical and cognitive interpretations that combine personal and social factors.

For example, if I dream that I have forgotten my suitcase on a train, the dream is not really about the particular scene depicted; rather, the dream relates together past experiences of forgetting important things and past experiences on trains. Furthermore, as Lacan insists, the unconscious is ethical because my sense of loss is affected by not only my personal sense of responsibility but also the way my society defines and punishes a lack of responsibility.[30] In terms of emotional responses to welfare policies, we should also consider that each individual has a particular set of associations, experiences, and interpretations related to the ethical judgment about sharing resources, and these emotional responses are both social and personal; in fact, from a psychoanalytic perspective, it is impossible to simply separate the social and the personal just as it is impossible to separate the emotional from the ethical and the cognitive.

Another psychoanalytic challenge to the way that emotions are defined by evolutionary psychology, neuroscience, and behavioral economics is in the difference between instincts and drives. For psychoanalysis, our inherited biological instincts are subverted and perverted by human sexuality because unlike other animals, humans can easily replace the objects, aims, and sources of instinctual pleasure.[31] Also, since humans can find pleasure in pain, there is a subversion of the idea that our instincts and emotions are derived from our ancestor's past successful attempts to resolve particular social problems.[32] As we saw in the case of Damasio, evolution has been misinterpreted as resulting in the sole purpose of individual survival through the successful adaption to a particular cultural and natural environment. I have argued that this scientific theory is in part a political ideology based on the neoliberal stress on how individuals conform to the social status quo. Yet, psychoanalysis helps us to understand that humans are not driven by the pure pursuit of self-regulation, adaptation, and individual survival. The psychoanalytic theories of the unconscious and the drives, therefore, disrupt both biological determinism and social determinism; human beings are prone to turn pleasure into pain as they reinterpret social norms through personal experiences and unconscious associations.

Contempt for the Other

This psychoanalytic understanding of human emotions stands in stark contrast to the way emotions are often defined by evolutionary psychologists, which is evident from the following passage:

> the same neural circuits are activated in moral disgust as in disgust for contaminating matter (e.g., blood, excrement) and, in parallel, both kinds of disgust motivate avoidance of further contact with its target (Rozin, Haidt, & McCauley, 2000). In contrast to both anger and disgust, contempt operates primarily in the domain of status, and the elicitation of contempt functions to communicate that the target is of lower status and to facilitate avoidance across levels of social hierarchies (Rozin et al., 1999).[33]

These evolutionary psychologists turn to neuroscience in order to argue that human emotions can be easily defined and connected to particular evolutionary goals that result in the inheritance of specific brain functions.[34] For instance, they argue that the mental contempt program is triggered when one encounters someone of a lower social status. Here, there is a narrow definition of an emotion that is then tied to a very specific social situation. In contrast, psychoanalysis shows that contempt can be shaped by a whole series of irrational, cultural, and personal interpretations that may go against a universal emotional response. While we may be able to agree that all humans have the ability to experience contempt, it seems absurd to argue that contempt always revolves around dealing with someone with a lower status, and even if we accepted this definition, the question remains of how a particular person determines that another particular person is of a lower status.

Like so many of the evolutionary psychology theories, this interpretation of contempt appears to be developed out of the need of the researchers to explain why people react to social welfare programs in a universal way that is determined by natural selection. As we shall see, much of this science is the result of reverse engineering where a present problem is explained by recreating an imagined past and then universalizing the results of the successful adaptation to a particular former conflict.[35] In this strange retrospective narrative, the feelings and responses of people existing hundreds of thousands of years ago are determined by projecting current values and beliefs into the unknowable distant past. Thus, the evolutionary psychologists have to imagine what problems were faced by our hunter-gatherer ancestors, and then they have to come up with how these humans responded to their particular social problems in order to survive their particular environments.[36] The trick here is that since evolution has no purpose or plan, a theory of successful adaptation has to be invoked in order to determine biological necessity.

Once the invented past is defined and its purposes are articulated, the next step is to show that we are currently using old evolutionary programs in a new environment:

> We suggest that many modern political issues, such as welfare, tax payments, criminal sanctions, redistribution, revolution, immigration, and race relations contain basic dilemmas that our ancestors evolved to deal with in order to successfully navigate social relationships (Alford & Hibbing, 2004; Cosmides & Tooby, 2006; Petersen, 2009, 2012). By implication, the mind of modern citizens is endowed with a toolbox of specialized mechanisms, including the social emotions, that could assist and facilitate their political decision making. Yet, these cognitive and emotion programs were designed in ancestral environments to respond to cues that were predictive in those environments.[37]

If I am right in showing that the invention of the past is used to justify behavior in the present, we should ask why is it necessary for evolutionary psychologists to make the argument that our current reactions to welfare state policies are largely shaped by the way our hunter-gatherer ancestors responded to their own particular social conflicts?

Naturalizing Neoliberalism

As we saw in the analysis of Pinker's work, part of the mission of evolutionary psychology is to discredit the social sciences and humanities because these other academic disciplines are not seen as being scientific, and one reason for this effort might be to enhance the ability of the new brain sciences to receive funding in the competitive grant system. After all, if they can prove that they have the answer and the other disciplines are just faulty speculations, then they will receive more research money and tenured professorships. I have also argued that behind this supposedly scientific discourse, we find a naturalization of the neoliberal status quo through the rationalization of competitive individualism. By stressing dire situations of survival, like the world faced by our ancestors, a call for individuals to focus on their own survival is reinforced. This argument also necessitates repressing psychoanalysis because the human focus on self-regulation and survival is challenged by the psychoanalytic theories of the unconscious and the drives.

Since our culture tends to value science above all other disciplines, the prestige and power tied to a scientific understanding of human nature is immense, and to protect this power of science, it is necessary to detach this discourse from anything that looks like political ideology or special interest.[38] Thus, for this science to be shown to be universal, neutral, objective, unbiased, non-ideological, and natural, it is necessary to discredit the importance of culture and language in shaping human behavior and thinking. In fact, one of the ways that these evolutionary psychologists try to show that their conception of the hunter-gatherer culture is scientific and not just a product of their imagination is by looking at the few remaining forager societies:

> Anthropological studies of living foragers have uncovered complex and pervasive systems of sharing both within and between families (Kaplan & Gurven, 2005). Taken together, hunter-gatherer studies, paleoanthropological evidence, and primatological evidence support the view that our ancestors have been engaging in social exchange for hundreds of thousands or millions of years (Cosmides & Tooby, 1992).[39]

The first problem with this use of current foragers to understand past hunter-gatherers is that the cultures of today are shaped by a very different natural and social environment, and so any direct comparisons are hard to make. It is also difficult to look at different animals and draw conclusions about how human beings think about particular social arrangements, and yet it is very common in the new brain sciences to use research on apes and other animals to understand how the human brain is wired.[40]

In the case of evolutionary psychology, it is essential to look at the distant past in order to comprehend how our minds currently function because the theory of evolution requires a long period of time to sort through the mechanisms that survive:

> In understanding the structure of the human mind and its constituent mechanisms, ancestral conditions are emphasized because the majority of our species only gave up forager life a few thousand years ago (Rindos, 1987). This is too little time for selection to engineer complex species-typical adaptations to novel postforager conditions ... Hence, whatever species-typical psychological mechanisms exist evolved in response to life in the preagricultural environment.[41]

By examining a very long historical record, the idea is to locate universal processes that transcend individual genetic variations. Since from this perspective, only the genetically determined universal mechanisms that ensure survival will last over time, the researchers have to connect their analysis of the current culture to the distant past; however, their theory of genetic evolution has to downplay the importance of mutations, recombinations, and environmental transformations. In fact, it is hard to say how their theory of evolution really fits with recent discoveries in genetics since these evolutionary psychologists appear to rely on the idea that our thoughts can be traced back to particular genetic material.[42]

Just as they apply a faulty theory of genetic evolution, they also impose a very narrow understanding of how our ancestors lived and survived:

> Our adaptations for social exchange evolved to operate in a world of small foraging groups. Importantly, the foraging niche that ancestral humans occupied involved the exploitation of large game, a high-quality, nutrient-dense, large-packaged food resource (Kaplan, Hill, Lancaster, & Hurtado, 2000). Such resources are difficult to acquire, and, hence, hunter-gatherers regularly experienced high variance in hunting success due to chance, illness, or other adversity (Hill & Hawkes, 1983; O'Connell, Hawkes, & Jones, 1991; Sugiyama, 2004). Such interruptions in the flow of calories posed an acute adaptive problem for our ancestors (Kaplan et al., 2000). At the same time, hunting successes would often provide more nutrients than a single individual or his family could consume at one time. While the capacity to store excess food was sharply limited, this enabled storage in the form of sharing and through the imposition of reciprocal sharing obligations from those one had shared with (Cosmides & Tooby, 1992; Lee & DeVore, 1968).[43]

This understanding of the hunter-gatherer culture is plausible, but is it scientific? Do we really know how these ancestors thought about their life possibilities and the way their groups distributed resources? Do we really know if all of these groups had the same experiences and social structures?[44] What we do know is that there were different environments and these environments changed over time, and if survival of the fittest means anything, it means the ability of specific genetic material to sustain itself in a particular environment for a particular amount of time.

Since these researchers believe that our ancestors must have been constantly affected by the inconsistent availability of food, they turn to behavioral economics to model how human and non-human behavior is shaped by varying resources:

> Consistent with this, a number of studies have shown that human foragers (and, perhaps, nonhuman primates) share a resource to a greater extent if the acquisition of the resource is subject to random variance ... Even for those who have more at any one time, such redistributive strategies would be adaptive if reversals of condition occur with sufficient frequency, and if an exchange is indeed reciprocal, i.e., if over the long run those who receive also give (Cosmides & Tooby, 1992; Trivers, 1971).[45]

Here we see how a modern understanding of behavioral economics is projected back into the hunter-gatherer culture and the way primates deal with scarce resources. An important move here is to assume that these social groups were always dealing with scarce resources, and so a universal solution had to be developed in order to make sure that a consistent and successful system of sharing was developed. Once again, this understanding of the past is freighted with assumptions that may be shaped by current political beliefs and values.

Since these researchers believe that our ancestors must have based their social interactions on the need to fairly distribute scarce and inconsistent resources, it is posited that it was necessary to develop a way of punishing people who appear to be cheating the resource distribution process: "One significant challenge, in the latter respect, is that practicing sharing exposes sharers to opportunistic exploitation by those that reap the benefits of others' productive efforts without incurring the costs of contributing (here called *cheaters*)."[46] By labeling these uncooperative hunter-gatherers "cheaters," these evolutionary psychologists reveal how their scientific investigation is influenced by the neoliberal conservative backlash against the postmodern welfare state. After all, the selection of the term "cheaters" is not innocent, and it reveals that one of the driving forces behind this study is to naturalize and rationalize the right-wing attack on people who receive social benefits. There are many other words that could have been chosen to define the people who did not share or give back enough resources, but the term cheaters is associated with "welfare cheats." Furthermore, this analysis relies on a simple binary opposition between cooperators and cheaters, which is a common way that behavioral economists examine economic exchanges through different types of experimental games like the Prisoner's Dilemma. However, a problem with this type of analysis is that it greatly reduces the variables that usually occur in human social interaction.[47]

Neoliberal Science

In order to show how all humans have inherited a mental mechanism that allows them to detect who is cheating the social system, the researchers look at previous research dealing with this topic:

> This entails the prediction that the human mind is well-adapted to detecting and reaction against cheaters. And as predicted, 30 years of experiments have shown that the human mind includes reasoning specializations for detecting cheaters in social exchanges, i.e., for solving tasks involving the identification of individuals who takes benefits without paying the required costs (Cosmides & Tooby, 2005). In support of the view that these specializations are adaptations, researchers have demonstrated their existence cross-culturally, including in small-scale societies (Sugiyama, Tooby, & Cosmides, 2002); that performance appears just as good in culturally unfamiliar as culturally familiar contexts (Cosmides & Tooby, 2005); that capacities for cheater detection is found in very young children (Harris, Nunez, & Brett, 2001); and that such capacities have distinct neural underpinnings (Stone, Cosmides, Tooby, Kroll, & Knight, 2002).[48]

It is interesting that they point to 30 years of research on detecting cheaters, and we can equate this time period with the neoliberal conservative backlash against social programs. Since the Age of Reagan, the general culture has internalized the notion that many people receiving government assistance are actually cheating the system because they are just lazy and have decided to live off government programs funded by the taxes of mostly wealthy people.[49] Yet, to prove that their research is not biased by this ideological perspective, they argue that the detection of welfare cheats can be found in people from very different cultures, and even young children have this natural ability to show contempt for welfare cheats. In fact, they insist that neuroscience has provided proof for the neural foundation for this social attitude.

As these evolutionary psychologists make this case of the natural reaction to lazy people who feed off of governmental programs, they are also forced to argue that there must be a natural instinct to share and cooperate:

Importantly, the detection of cheaters also has the predicted behavioral effects. Hence, evidence from experimental economic games shows that people cease to contribute to a public good if others don't follow suit (see, e.g., Fehr & Gächter, 2000). Similarly, observational studies of foragers (Kaplan & Gurven, 2005; Kaplan & Hill, 1985) indicate that food sharing among nonkin is to a significant degree reciprocal, i.e., conditional in the sense that A shares with B, if B shares with A.[50]

According to the logic expressed above, this research could have been focused on why we have evolved to share and cooperate; in fact, they appear to be making a strong claim for some form of communism, and yet, they have decided on focusing their attention of why people do not support social welfare programs.

To prove that people have an inherited program dedicated to detecting welfare cheaters, the authors stress the need for individuals to be able to determine if other people are being lazy: "In discriminating between cheaters and noncheaters, evolutionary analysis suggests that our minds have been designed to especially attend to cues of others' motivation to take part in the system of social exchange, i.e., their willingness to accrue and exchange resources."[51] The question remains how does one determine the level of another person's ability to work and provide resources and how does one judge reciprocity? Moreover, in a mass society, how do people judge the work ethics of millions of other individuals?

On one level, this research claims that our inherited cheater detector is always at play when we think about social welfare programs, but our individual responses are shaped by how we interpret relevant social information, and this inclusion of cultural knowledge threatens to undermine the importance of the universal response mechanism. For instance, just because all people can feel anxious, it does not mean that people experience anxiety for the same reasons; however, the tendency for evolutionary psychologists is to take the next step and search out universal situations matched with universal psychological reactions.[52] Therefore, when they discuss the cheater detector program, they specify what triggers these reactions on a universal basis:

> In line with these arguments, experimental studies show that humans represent bad outcomes with different mental categories depending on whether those outcomes are attributable to incompetence or lack of motivation (Delton et al., 2012). Similarly, a range of studies in neuroscience

have provided evidence for the important role of intentions in cheater detection (for an overview of this research, see Petersen, Roepstorff, & Serritzlew, 2009). For example, fMRI studies have demonstrated that cheater-detection tasks engage distinct theory-of-mind-related neural circuits (i.e., circuits involved in gauging the intentions of others), which are not engaged by other logically equivalent tasks (see Ermer, Guerin, Cosmides, Tooby, & Miller, 2006).[53]

In turning to neuroscience, evolutionary psychology seeks to enhance the scientific aura of its research, and yet it still has to deal with the problem of defining particular emotional programs and how they relate to specific social situations. Thus, they argue here that functional magentic resonance imaging (fMRI) studies show that when people use their cheater detectors to react to receivers of social welfare, a particular neural network is activated, and this network has been associated with interpreting the intentions of others. This use of neuroscience brings up several issues since it assumes that a single brain region deals with judging the thinking of others, and this activity can be clearly registered in a brain imaging technology.[54] There is also the problem of whether most people really think about the intentions of the receivers of benefits when they respond to social policies. From a universalizing democratic perspective, many people believe that it is simply fair and just to help out all people in need, and there is no reason to make a moral judgment regarding the intentions of the potential receivers of aid.[55]

For these evolutionary psychologists, it is not only important to show that we have inherited an automatic response to welfare cheaters, but we are also preprogrammed to punish the people we think are cheating the system:

In the face of cheaters (individuals with parasitic motivations), however, the adaptive response entails avoiding sharing but also attempts to recalibrate the cheater's motivational system to be more cooperative. The latter is important. In ancestral small-scale groups, there would be only a limited number of potential valuable social relationships, and hence it would be important not just to dismiss strategic cheaters but to recalibrate them so that they become better cooperators. This strongly suggests that the detection of cheaters in a sharing situation should trigger anger—the emotion designed to defend against exploitation and incentivize the up-regulation of investments by others (e.g., Sell et al., 2009)—rather than the more avoidance-oriented emotions of contempt and disgust ... More generally, cross-cultural studies of everyday morality show that anger and compassion are in fact regulated by effort-related perceptions in the face of needy individuals. Hence, Weiner (1995) reports studies from the United States, Canada, and Japan that consistently show that subjects respond with high levels of anger and low levels of compassion to a lack of effort among individuals requesting help.[56]

This cross-cultural analysis of the role played by anger in trying to train cheaters not to exploit social sharing introduces another set of problems. Since it is important for the authors to argue that these psychological reactions and social interactions are universal and natural, they have to discount cultural influences and cultural differences. Once again, from a psychoanalytic perspective, it is vital to show how cultural influences are internalized through identification, and this process of personalization often occurs without conscious awareness. Universality and biology are thus disrupted by the unconscious and the preconscious as social representations are internalized and misrepresented.

The Biological Unconscious

However, to affirm the universal and biological nature of popular responses to welfare programs, the authors make the following 14 predictions:

1. Feelings of compassion toward welfare recipients increase support for welfare.
2. Feelings of anger, contempt, and disgust toward welfare recipients decrease support for welfare.
3. Feelings of anxiety when thinking about welfare recipients have little or no effect on support for welfare.
4. Feelings of anger, contempt, and disgust to welfare recipients make opinion formation on welfare issues faster.
5. Feelings of compassion for welfare recipients make opinion formation on welfare issues faster.
6. Feelings of anxiety when thinking about welfare recipients do not make opinion formation on welfare issues faster.
7. Ideological predisposition does not influence the speed with which individuals form opinions on welfare issues.
8. Welfare recipients with little motivation to look for work elicit anger.
9. Welfare recipients motivated to look for work elicit compassion.

10. The activation of anger partially mediates the effect of effort cues on support for welfare.
11. The activation of compassion partially mediates the effect of effort cues on support for welfare.
12. Compassion and anger mediate the opinion effects of effort cues independently of political ideology.
13. Cues of effort regulate anger (and compassion) rather than anger-related emotions such as anxiety, contempt, and disgust.
14. Anger and compassion are regulated by cues of effort rather than by effort-related cues of competence.[57]

A key element of these predictions is the idea that only certain emotions are social, and if a reaction is fast, it must be automatic and therefore derived from evolution. In other words, following the tradition in neuroscience and behavioral economics to divide the mind into an automatic system and a reflective system, the argument here is that any speedy emotional response must be instinctual, non-conscious, intuitive, biological, and preprogrammed by evolution. Of course, what this theory leaves out is the possibility that some fast responses might be due to social prejudices, internalized ideologies, or unconscious associations.[58]

In order to test whether responses to welfare programs are automatic and thus derived from evolution, the studies focus on how quickly people respond to particular questions.[59] Once again, the assumption is that fast responses indicate unconscious and "evolved" mechanisms:

> Converging results from several areas of research suggests that evolved dedicated circuits tend to operate more rapidly and effortlessly within their natural domain than do acquired skills or dedicated circuits used outside their domain (Ermer et al., 2006; New, Cosmides, & Tooby, 2007). For this reason, we expect social emotions will influence welfare opinions in ways that are distinct from other emotions and opinion factors.[60]

In the passage above, the stress on "circuits" reflects the notion derived from neuroscience that regions of the brain function like computers preprogrammed by evolution.[61] Furthermore, the authors equate fast responses with an absence of ideology or opinion, and therefore, they indicate that strong, quick rejections of welfare programs must be the result of biology and not culture, experience, or politics. In other words, they tend to argue that while some people's perceptions can be shaped by ideology and culture, the people who have an immediate reaction are responding on a purely intuitive and biological level. This theory indicates that the neoliberal conservative backlash is driven by evolution and not politics.[62] What is so problematic about this claim is that the right-wing ideology is seen as being non-ideological as the role played by unconscious association is repressed.

Although in the early part of their study they indicate that they do not dismiss ideology or culture, it is clear that their goal is to show that these social factors do not really matter:

> we expect compassion and anger to mediate effort cues independently of political ideology ... if it is the logic of social exchange, built into the structure of the anger and compassion systems, that makes modern citizens sensitive to effort cues, we should expect their operations to be independent of the effects of political ideology. Conservatives and liberals are expected to share the same species-typical mental architecture and, hence, should be angered or feel compassionate by exposure to the same ancestrally relevant cues. That is, while conservatives and liberals disagree in the abstract about welfare recipients, these general differences should drop in importance when individuals across the political spectrum are provided with the same cues, as long as these cues fit the input conditions of the emotional programs (for a more extended discussion, see Petersen, 2009).[63]

These researchers do not consider that one reason why liberals and conservative may be responding in the same way is that the neoliberal conservative ideology has been internalized by both liberals and conservatives: after hearing the message for 40 years about the need to eliminate welfare benefits, the general populace has internalized on an unconscious and intuitive level this backlash discourse.

Since neoliberalism is in part based on the idea that we cannot afford to help out the poor, and so welfare programs have to be reduced, it has been necessary to demonize the benefactors of certain government programs, and this demonization has fueled a rhetoric of anger and resentment.[64] In the case of neuroliberalism, this neoliberal ideology is given a neurobiological foundation: "First, the argument that emotional sensitivity to effort cues reflects adaptive strategies designed to facilitate investments in reciprocal social exchanges allowed us to pinpoint anger as the distinctive aversive emotion triggered in the face of lack-of-effort individuals requesting help. Feather (2006, p. 46) argues that lack of effort triggers feelings of resentment."[65] Due to their focus on anger and resentment, the authors argue that these immediate responses must be the result of the ways our ancestors responded to people who did not seem to work for their share of common resources. What is neglected here is the historical periods and cultures that have supported a large welfare state. For instance, before the Age of Reagan, there was a 40-year period in the United States during which New Deal social welfare programs were rarely called into question; there are also many countries in Northern Europe that still support universal welfare programs that do not seek to make moral judgments about the recipients.[66]

The researchers' hidden bias against social welfare programs can be detected in the questions they ask in one of their studies:

> "High incomes should be taxed more than is currently the case" (reverse coded), "We should resist the demands for higher welfare benefits from people with low incomes," "The wealthy should give more money to those who are worst off" (reverse coded), "The government spends too much money on the unemployed," "The state has too little control over the business world" (reverse coded), and "In politics, one should strive to assure the same economic conditions for everyone, regardless of education and employment" (reverse coded).[67]

One major problem with these survey questions is that they mostly pit the wealthy against the poor, and they feed into the neoliberal idea that rich people have to pay taxes to support low-income people, but the reality is that governments spend money on many different things, and welfare is often just one part of a country's budget. Yet, the neoliberal conservative backlash has tried to equate all government programs with welfare and all taxes with the poor stealing from the wealthy.[68] Therefore, by asking people to respond to these ideological questions, the responses are influenced by the unconscious assumptions of neoliberal politics.

After the researchers discovered that ideology did have a major effect on how people responded to issues concerning welfare programs, they interpret the results in the following manner:

> While both social emotions and ideology predict welfare opinions, the evolutionary perspective suggests that they operate in different ways in the opinion-formation process. Social emotions are evolved psychological systems designed to facilitate social responses that would have been adaptive ancestrally. In contrast, the ideologies of left and right are relatively recent, culturally elaborated constructs whose complex specifics must be memorized.

These two processes of opinion formation (triggering existing circuits versus reasoning from memorized data structures) are fundamentally different in kind. Thus, Predictions 4–7 entail that social emotions not only shape the content of subjects' welfare opinions but (unlike ideology) the more they are activated, the more rapidly they should organize welfare opinions. Thus, increasing intensity of emotional reactions to welfare recipients should be associated with lower response times when answering questions about welfare opinions. Ideology should not. To test this prediction, Model 3 in Table 2 regresses subjects' response times to welfare opinions on the different explanatory variables.[69]

In plain English, they argue that the fact that the people with the fastest responses had the strongest responses must mean that their feelings were automatic and biological: "As predicted, we see significant negative effects of the two social emotions scales on response times. The more intensely subjects feel aversive or compassionate emotions towards welfare recipients, the faster they respond to questions about welfare. In contrast, there are no effects of either anxiety or ideology on response time."[70] Here we enter a self-confirming loop since they have already made the assumption that biology and not ideology results in fast responses. So even if someone has internalized a political perspective and is able to make quick responses to social cues regarding that ideology, the researchers still consider that the responses are not ideological.

It should be obvious that there can be many causes for the speed of a response to an online survey, and yet much of this study is centered on the reductive argument that the fastest responses can only be derived from inherited mental programs.[71] Moreover, the authors openly dismiss the importance of social learning, culture, and politics by arguing that these social influences are relatively new human constructs and thus cannot be considered to be as important as mental programs inherited from natural selection. These researchers also argue that social welfare programs are inherently a Leftist project and not policies that have ever had any bipartisan support: "One could, for example, have argued that the welfare state has been the central political project of the left, and, therefore, those belonging to the ideological left should be able to answer questions about welfare rapidly. As revealed in the model, this is not the case."[72] What the authors do not consider here is that people supporting welfare programs might be more reflective of their social opinions, while the people rejecting these programs might be less reflective and more affected by internalized political ideology. From the perspective of the researchers, it is inherited morality and not learned ideology that determines the speed and intensity of responses: "While ideology is a powerful predictor of welfare opinions, citizens do not make inferences from ideology with the same ease as they do from their moral feelings."[73] This interpretation pushes us to ask how do the authors differentiate ideology and morality, and is it possible to have moral responses that are also not ideological?

From a psychoanalytic perspective, unconscious feelings and responses are shaped by both personal and social associations, and this means that it is hard to separate ideology from morality. For instance, according to Freud, dreams use distorted symbolism to circumvent an internalized censor, and this censor must be derived from morality, which itself is shaped by ideology.[74] What the evolutionary psychologists do is to eliminate the importance of psychoanalysis by arguing that our unconscious is not due to repression or the avoidance of censorship; rather, they see unconscious responses as simply preprogrammed by natural selection.

Punishing Welfare Snakes

After performing several studies that look at how people judge differently lazy versus incompetent recipients of welfare, the researchers make the following judgment:

> Evolutionary psychologists and hunter-gatherer researchers have developed multiple, converging lines of evidence that support the view that social exchange functioned among our ancestors as a social insurance strategy through which individuals could guard against interruptions in the food supply due to injury or bad luck. By sharing with others, individuals invest in future help. By making that sharing conditional on whether potential recipients were disposed to contribute when they could, sharers protected themselves against exploitation. The recurrent payoffs to conditional cooperation—extending over hundreds of thousands of years—selected for psychological mechanisms in our species that reliably guided our ancestors to implement this winning strategy. On this view, the social emotions of anger and compassion evolved, in part, to motivate these investment decisions.[75]

Once again, the language used in this seemingly scientific, objective analysis reveals a set of ideological biases. On one level, they argue that it is the people with the most resources who are "exploited" by the people with the least resources, and on another level, they declare it a "winning" strategy to punish cheaters and people who appear to be unmotivated.

Furthermore, in an effort to naturalize social ideologies, the authors compare punishing welfare cheats with the inherited fear of snakes:

> To risk oversimplification, just as we have evolved specializations that cause us to fear snakes and spiders, we evolved specializations that make us angry at the lazy but compassionate toward the needy. In this way, with evolved cognitive and emotion programs as the intermediate link, the configuration of past adaptive problems is responsible for structuring aspects of public opinion on present political issues in a way that has not been widely appreciated. That is, public opinion turns out to be sensitive to cues that were relevant for social navigation in ancestral small groups—even though these cues might not be important (or may even be counterproductive) to respond to in modern societies. In the case of welfare, it is the perception of welfare recipients' motivation to work, and not their competence, that is the more powerful determinant of welfare opinion.[76]

This interpretation is clearly conservative in the sense that it insists that our current political responses to welfare programs are the result of unchangeable mental programs derived from evolution.

As part of the neoliberal conservative backlash against progressive parenting, education, and politics, the message here is clear: we cannot overcome thousands of years of evolution through conscious social interventions to make society more just and fair. In other words, liberalism is dead because you cannot change human nature since this nature is biological and inevitable, and yet, toward the end of this research paper, the authors appear to bring back culture and ideology back into the discussion:

> However, while individual and cross-national differences do not change the structure of the emotional mechanisms, we do not in any way intend to say that such differences are unimportant (see, e.g., Alford, Funk, & Hibbing, 2008; Jost, Nosek, & Gosling, 2010). Rather, we want to emphasize that their importance rests in providing input to these and other mechanisms in the absence of any externally provided and vivid cues about, for example, specific welfare recipients. In such situations, we should expect our emotional systems to fall back on extracting the cues necessary for their execution from internally provided perceptions, images, and stereotypes about the motivations of welfare recipients.[77]

In this final gesture toward the importance of culture and ideology, we encounter a major contradiction: if people in mass society today rarely are making judgments about social welfare programs based on immediate encounters with other people, does this mean that they are always being influenced by stereotypes and other social influences? If this is true, does not this call into question the value of the entire attempt to base responses to welfare programs on inherited mental programs? In other words, if we only encounter welfare recipients through media images and stereotypes, does that not mean that our inherited programs are always triggered by cultural content, and therefore, it is culture that determines our responses?[78]

From a psychoanalytic perspective, this contradiction represents a symptom that reveals how the effort to prove the universality of biological human nature has often been a political project dedicated to undermining progressive social interventions. Meanwhile, as stereotypes are naturalized, prejudices are rationalized, and the underlying ideology of the neuroliberal backlash is exposed. For example, in an interesting moment near the end of their research paper, the authors finally introduce the issue of race in shaping how people respond to social welfare programs:

> Denmark is relatively homogeneous, and neither race nor ethnicity plays a major role in discussions on social welfare (Larsen, 2006). In contrast, the United States is highly heterogeneous and, in addition, a disproportionately large number of black Americans are on social welfare (Alesina et al., 2001). By implication, race plays a key role in opinions on social welfare in the United States, and white Americans' opposition to social welfare seems to be driven primarily by the perception that black Americans are lazy (Gilens, 1999). This difference provides an illustration of how macrostructural conditions shape our behaviors through a number of interacting psychological mechanisms. Previous research in evolutionary psychology suggests that the human mind does not include a dedicated system for categorizing by race.[79]

Here we see how political ideology, prejudice, and cultural stereotypes do shape how people think about social welfare programs, and so it is important to ask what do evolutionary psychology, neuroscience, and behavioral economics actually bring to our understanding of this issue and other social concerns? If our inherited mental programs still rely on cultural information to shape responses, what is gained by knowing that some response mechanisms are shaped by evolutionary forces? I have been arguing that these new brain sciences perform a mostly negative function of discrediting other disciplines and non-biological factors as they naturalize the neoliberal social status quo.

NOTES

1. Harding, Sandra. "After the Neutrality Ideal: Science, Politics, and "Strong Objectivity"." *Social research* (1992): 567–587.
2. Petersen, Michael Bang, et al. "Who deserves help? evolutionary psychology, social emotions, and public opinion about welfare." Political psychology 33.3 (2012): 395.
3. Haidt, Jonathan. *The righteous mind: Why good people are divided by politics and religion.* Vintage, 2012.
4. Peterson et al., 395.
5. Hancock, Ange-Marie. *The politics of disgust: The public identity of the welfare queen.* NYU Press, 2004.
6. Peterson et al., 395–6.
7. Ibid., 396.
8. Coughlin, Richard M. *Ideology, public opinion, & welfare policy: attitudes toward taxes and spending in industrialized societies.* No. 42. Univ of California Intl &, 1980.
9. Shepherdson, Charles. *Vital signs: Nature, culture, psychoanalysis.* Psychology Press, 2000.
10. Freud, Sigmund, and James Strachey. *Three essays on the theory of sexuality.* Vol. 5008. Basic Books, 1975.
11. Lacan, Jacques. "The mirror stage as formative of the function of the I as revealed in psychoanalytic experience." *Cultural Theory and Popular Culture. A Reader* (1949): 287–292.
12. Peterson et al., 396.
13. Longobardi, Giuseppe, and Ian Roberts. "Universals, diversity and change in the science of language: Reaction to "The Myth of Language Universals and Cognitive Science"." *Lingua* 120.12 (2010): 2699–2703.
14. Peterson et al., 396.
15. Sampson, Edward E. "Cognitive psychology as ideology." *American psychologist* 36.7 (1981): 730.
16. Rose, Steven. "Alas, poor Darwin." *Biologist (London, England)* 48.2 (2001): 100.
17. Peterson et al., 396.
18. Ibid, 397.
19. Ketelaar, Timothy, and Bruce J. Ellis. "Are evolutionary explanations unfalsifiable? Evolutionary psychology and the Lakatosian philosophy of science." *Psychological Inquiry* 11.1 (2000): 1–21.
20. Holcomb III, Harmon R. "Just so stories and inference to the best explanation in evolutionary psychology." *Minds and Machines* 6.4 (1996): 525–540.
21. Peterson et al., 397.
22. Kashtan, Nadav, Elad Noor, and Uri Alon. "Varying environments can speed up evolution." *Proceedings of the National Academy of Sciences* 104.34 (2007): 13711–13716.

23. Peterson et al., 397.
24. Ibid.
25. Rose, Hilary, and Steven Rose. *Genes, cells and brains: The promethean promises of the new biology.* Verso Books, 2013.
26. Ibid., 397–8.
27. Freud, Sigmund, Marie Ed Bonaparte, Anna Ed Freud, Ernst Ed Kris, Eric Trans Mosbacher, and James Trans Strachey. "Project for a scientific psychology." (1954).
28. Freud, Sigmund. *The interpretation of dreams.* Read Books Ltd, 2013.
29. Lacan, Jacques. *The Ego in Freud's Theory and in the Technique of Psychoanalysis, 1954–1955.* Vol. 2. WW Norton & Company, 1991.
30. Lacan, Jacques. "The Four Fundamental Concepts of Psychoanalysis, trans. Alan Sheridan." *New York: Norton* 67 (1978): 73–77.
31. Freud, Sigmund. "Instincts and their vicissitudes." *The Standard Edition of the Complete Psychological Works of Sigmund Freud, Volume XIV (1914–1916): On the History of the Psycho-Analytic Movement, Papers on Metapsychology and Other Works.* 1957. 109–140.
32. Freud, Sigmund. "The economic problem of masochism." *The Psychoanalytic Review (1913–1957)* 16 (1929): 209.
33. Peterson et al., 398.
34. Panksepp, Jaak, et al. "Comparative approaches in evolutionary psychology: Molecular neuroscience meets the mind." *Neuroendocrinology Letters* 23.Suppl 4 (2002): 105–115.
35. Peters, Brad M. "Evolutionary psychology: neglecting neurobiology in defining the mind." *Theory & Psychology* 23.3 (2013): 305–322.
36. Smith, Barbara Herrnstein. "Sewing up the mind: the claims of evolutionary psychology." *Alas, poor Darwin: Arguments against evolutionary psychology* (2000): 129–143.
37. Peterson et al., 398.
38. Meyer, John W., and Ronald L. Jepperson. "The 'actors' of modern society: The cultural construction of social agency." *Sociological theory* 18.1 (2000): 100–120.
39. Peterson et al., 399.
40. Byrne, Richard, and Andrew Whiten. "Machiavellian intelligence: social expertise and the evolution of intellect in monkeys, apes, and humans (oxford science publications)." (1989).
41. Peterson et al., 398.
42. Rose, Hilary, and Steven Rose. *Genes, cells and brains: The promethean promises of the new biology.* Verso Books, 2013.
43. Peterson et al., 399.
44. Ingold, Tim. "The optimal forager and economic man." *Nature and society: Anthropological perspectives* (1996): 25–44.

45. Peterson et al., 399.
46. Ibid.
47. Green, Donald P., Ian Shapiro, and Ian Shapiro. *Pathologies of rational choice theory: A critique of applications in political science.* New Haven: Yale University Press, 1994.
48. Peterson et al., 399–400.
49. Martin, Isaac William. *The permanent tax revolt: How the property tax transformed American politics.* Stanford: Stanford University Press, 2008.
50. Peterson et al., 400.
51. Ibid.
52. Confer, Jaime C., et al. "Evolutionary psychology: Controversies, questions, prospects, and limitations." *American Psychologist* 65.2 (2010): 110.
53. Peterson et al., 400.
54. Hickok, Gregory. *The myth of mirror neurons: The real neuroscience of communication and cognition.* WW Norton & Company, 2014.
55. Rawls, John. "Distributive justice." *Perspectives In Bus Ethics Sie 3E* (1967): 48.
56. Peterson et al., 400.
57. Ibid., 401.
58. Vedantam, Shankar. *The hidden brain: How our unconscious minds elect presidents, control markets, wage wars, and save our lives.* Random House Digital, Inc., 2010.
59. Kahneman, Daniel. *Thinking, fast and slow.* Macmillan, 2011.
60. Peterson et al., 402.
61. Innocenti, Giorno Maria. "Neuroscience, brains, and computers." *Pensamiento. Revista de Investigación e Información Filosófica* 67.254 (2011): 609–615.
62. Jimenez, Guillermo C. *Red Genes, Blue Genes: Exposing Political Irrationality.* Autonomedia, 2009.
63. Peterson et al., 402.
64. McCluskey, Martha T. "Efficiency and social citizenship: challenging the neoliberal attack on the welfare state." *Indiana Law Journal* 78 (2003).
65. Peterson et al., 402.
66. Hill, Steven. *Europe's Promise: Why the European way is the best hope in an insecure age.* Univ of California Press, 2010.
67. Peterson et al., 404.
68. Fraser, Nancy. "Clintonism, welfare, and the antisocial wage: the emergence of a neoliberal political imaginary." *Rethinking Marxism* 6.1 (1993): 9–23.
69. Peterson et al., 405–6.
70. Ibid., 406.

71. Truell, Allen D., James E. Bartlett, and Melody W. Alexander. "Response rate, speed, and completeness: A comparison of Internet-based and mail surveys." *Behavior Research Methods, Instruments, & Computers* 34.1 (2002): 46–49.
72. Peterson et al., 406.
73. Ibid.
74. Freud, Sigmund. "Some additional notes on dream-interpretation as a whole." *The Standard Edition of the Complete Psychological Works of Sigmund Freud, Volume XIX (1923–1925): The Ego and the Id and Other Works.* 1961. 123–138.
75. Peterson et al., 413.
76. Ibid.
77. Ibid.
78. Herman, Edward S., and Noam Chomsky. *Manufacturing consent: The political economy of the mass media.* Random House, 2010.
79. Peterson et al., 414.

CHAPTER 6

Drugging Discontent: Psychoanalysis, Drives, and the Governmental University Medical Pharmaceutical Complex (GUMP)

Abstract In this chapter, I place this conflict between the new brain sciences and psychoanalysis in a very particular context in order to provide a pressing example of the destructive effects of neuroliberalism. My central claim is that governments, universities, medical doctors, and pharmaceutical companies are *unintentionally* working together to position medication as the only solution to most psychological and social problems, and this focus on the use of drugs not only represses psychoanalysis but also relies on the basic argument of the new brain sciences. Through the formation of what I will be calling the Governmental University Medical Pharmaceutical Complex (GUMP for short), an unintentional collusion has been produced that transforms science into a neoliberal market as a new form of social Darwinism is circulated. Moreover, to offer a counterdiscourse to GUMP, I articulate how psychoanalysis represents a vital challenge to the drugging of discontent.

Keywords Brain sciences • Pharmacology • Psychoanalysis • Universities • Medicine • Neoliberalism

Throughout this book, I have turned to psychoanalysis to offer a political critique of the new brain sciences in the age of neoliberalism. My central claim has been that neuroscience, evolutionary psychology, and behavioral economics often represent a political project seeking to be viewed as a pure science. Since these new discourses tend to focus on how individuals

© The Author(s) 2017
R. Samuels, *Psychoanalyzing the Politics of the New Brain Sciences*,
https://doi.org/10.1007/978-3-319-71891-0_6

are programmed by evolution to pursue their own survival by adapting to the social world around them, neuroliberalism serves to naturalize inequality and the culture of competitive individualism. In fact, the theory of natural selection and the current neoliberal ideology of meritocracy share a lot in common since both systems argue that individuals compete for scarce resources, and the ones who are the most successful are rewarded. Both systems also rely on the notion that success is achieved by adapting to a specific environment and that there will naturally be winners and losers in a competitive market with limited resources. Here we see how the ideology of the free market shapes our conceptions of evolution and meritocracy because all of these systems are centered on the logic of the Invisible Hand and the notion that the common good will be generated through the actions of individual self-interest.

Naturalizing the Social

Karl Marx once remarked that when Darwin looked at nature, he saw contemporary England: "It is remarkable how Darwin recognizes among beasts and plants his English Society with its division of labor, competition, opening of new markets, 'inventions,' and the Malthusian 'struggle for existence.'"[1] In other words, from the start, the theory of evolution was shaped by the capitalist logic of competitive markets and social hierarchies. neuroliberalism, then, represents an extension of this market-based logic by seeing genes, neurons, and neurotransmitters as engaged in a competitive struggle for survival. For instance, in *The Selfish Gene*, Richard Dawkins argues that humans are shaped by the evolutionary drive to replicate genetic material, and these efforts at self-duplication set up a competition among the genes that are best suited to survive.[2] While in a meritocracy or capitalist market it is the self-interested individual who competes to survive, in evolution, it is genetic material that is determined to reproduce. Once again, in this new social Darwinism , we can see how scientific theory is shaped by contemporary ideology, and what I have been calling neuroliberalism.

According to the most basic argument of the new brain sciences, human thought and behavior is determined by inherited programs that have been produced through genetic replication of particular neural structures powered by identifiable neurotransmitters and localized in detectable brain regions. These inherited programs were developed during a distant past to solve particular problems, which aided our ancestors in adapting and surviving in a severe environment with limited resources. Although some neuroscientists, evolutionary psychologists, and behavioral economists reject different aspects of the basic argument I have articulated, it is hard to eliminate any part of this theory without the whole system falling apart. For example, if one wants to affirm that human behavior is shaped by inherited programs, one must be able to trace those programs to specific genes, neurons, and neurotransmitters. Moreover, if one turns to inherited genes to determine specific patterns of behavior, one has to employ an explanation for what has caused this genetic material to survive, and this leads one to posit that the inherited programs must have helped people survive in a particular environment, and since it takes a very long time for natural selection to weed out the losers, the inherited programs must come from the distant past. Moreover, each program would have to be tied to resolve successfully a particular problem that helped the genetic material to replicate.

In contrast to the basic argument of the neuroliberal brain sciences, I have argued that psychoanalysis presents several challenges. First of all, the unconscious and human sexuality disrupt the inherited mental programs by allowing for mental autonomy, chance, and contingency. Since on the level of the unconscious, anything can be a source of sexual satisfaction, including pain, then it is hard to imagine how human behavior is dedicated strictly to self-regulation and survival. Furthermore, due to the fact that humans are able to confuse fact with fantasy, they can misread their environment and their own actions and thoughts. Although psychoanalysis does recognize the power of inherited biological forces, it also affirms the various ways culture and subjectivity disrupt biological determinism. For example, Freud early on realized that hysterical symptoms did not make anatomical sense, and thus the body of the neurotic was not a purely empirical or biological entity.

In this chapter, I will place this conflict between the new brain sciences and psychoanalysis in a very particular context in order to provide a pressing example of the destructive effects of neuroliberalism. My central claim is that governments, universities, medical doctors, and pharmaceutical companies are *unintentionally* working together to position medication as the only solution to most psychological and social problems, and this focus on the use of drugs not only represses psychoanalysis but also relies on the basic argument of the new brain sciences. Through the formation of what I will be calling the Governmental University Medical Pharmaceutical Complex (GUMP for short), an unintentional collusion has been produced that transforms science into a neoliberal market as a new form of social Darwinism is circulated. Moreover, to offer a counter-discourse to GUMP, I articulate how psychoanalysis represents a vital challenge to the drugging of discontent.

THE BASIC ARGUMENT

Throughout this chapter, I will argue that in order for pharmaceutical companies to convince the public and government agencies to buy into seeing medication as the only valid treatment for psychological problems, they need the backing of a scientific explanation that discredits other types of treatment and explains why our disorders are derived from inherited mental programs. It is essential to stress here that the pharmaceutical corporations also require the support of the medical establishment, especially psychiatry, in order to promote their solution, and once again, this support relies on discrediting psychoanalysis and other types of psychotherapy in favor of a model of mental functioning based on biological determinism. Likewise, in order to test their medications and promote their ideology, Big Pharma also needs governmental agencies, like the National Institute of Mental Health, to fund university research to test drugs and produce biomedical theories and research.[3]

The GUMP complex is therefore not a coordinated conspiracy, but in actuality is the result of several different trends that we can trace back to neoliberal ideology. The most evident connecting element is the notion that all public services, including education, science, healthcare, and therapy, should be centered on making profits within a competitive market. Of course an obvious example of the dominance of neoliberal priorities is the pharmaceutical industry, which clearly focuses on making money and does not see their main goal as helping people live healthier lives.[4] Although Big Pharma does argue that they provide an important public good, they continue to pursue privatized profits. Likewise, psychiatrists may believe that they are doing the right thing, and medication is the only solution to most mental health problems, but their turn toward psychotropic drugs has often been motivated by a desire to increase their prestige and their wealth.[5] Moreover, the medical industry is now tightly controlled by the insurance companies that dictate what kind of treatment goes with what kind of mental disorder.[6] Furthermore, as a result of the neoliberal defunding of universities, these educational institutions have been forced to rely on outside grant funding, and a lot of the support comes from pharmaceutical corporations. In fact, the very process of funding scientific research out of external grant results in the possibility of research being corrupted by the need to please the funders to receive future support.[7]

GUMP AND MEDICALIZING MENTAL DISORDERS

As university researchers, psychiatrists, government agencies, and pharmaceutical corporations are all motivated to buy into the biological determinism of psychological behavior, these efforts are enhanced by the desire of individual patients to see their mental problems as a disease and not the result of social inequality or individual choice. Like the positioning of addiction as a disease, one of the effects of this understanding is that the affected individual is seen as a helpless victim.[8] Unlike psychoanalysis, where someone in therapy has to consider the moral, psychological, familial, and cultural sources for particular problems, the person given psychiatric medication is often represented as having no role to play in understanding or transforming his or her own behavior. Furthermore, as Freud discovered with psychosomatic illnesses, a defined disease gives someone a set identity and a way of retreating from the problems of the external world.[9] In fact, one of Freud's most controversial and important moves was his discovery that many of the scenes of victimization presented by his patients were actually the result of fantasies and the masochistic enjoyment of pain. Since the victim of an illness is always innocent and cannot be questioned, people gain a psychic reward from attaching their discontent and disorders to biological causes.

In *Anatomy of an Epidemic: Magic Bullets, Psychiatric Drugs, and the Astonishing Rise of Mental Illness in America*, Robert Whitaker documents how many psychotropic drugs have been sold to the public by equating psychological pathologies with diseases like diabetes.[10] Just as insulin is seen as the miracle cure for this physical disorder, Ritalin, Valium, and other psychotropics have been represented as the magical pills to cure specific mental disorders.[11] Whitaker's analysis of the reasons why we became a society of drug takers and drug pushers begins with his acknowledgement that he was once a major promoter in his journalism of the magic pill solution to mental health issues: "I believed that psychiatric researchers were discovering the biological causes of mental illnesses and that this knowledge had led to the development of a new generation of psychiatric drugs that helped 'balance' brain chemistry. These medications were like 'insulin for diabetes.' I believed that to be true because that is what I had been told by psychiatrists while writing for newspapers."[12] As Whitaker stresses throughout his work, a key enabling force behind the successful promotion of so many psychiatric drugs was an unproven scientific theory centered on the idea that mental health problems were just like medical diseases that could be cured or managed by a chemical solution. Since this story was so simple for the general public to understand, it was easy for the pharmaceutical companies to sell.

It is important to stress here that at times drug treatments of severe mental disorders do work, and so it is not a question of simply abandoning the medical treatment of psychological problems. Rather, my argument is that the chemical solution is overused and clinicians fail to distinguish between disorders caused by biology and psychical impairment to the brain on the one hand, and mental disorders that have a psychological foundation on the other. Moreover, since the psychoanalytic treatment and theory has been undermined by psychiatrists, evolutionary psychologists, and neuroscientists, the therapeutic alternative has been repressed.

According to Whitaker, not only did famous university scientists and doctors endorse the unproven theory of the biological causality of all mental disorders, but each new drug was cheered on by high-ranking governmental officials: "The introduction of Prozac and other 'second-generation' psychiatric drugs, the surgeon general added, was 'stoked by advances in both neurosciences and molecular biology' and represented yet another leap forward in the treatment of mental disorders."[13] It is important to stress here that in the United States, the National Institute of Mental Health and the National Institute of Health provided much of the funding to universities to articulate and circulate theories purporting to prove that chemical imbalances were the cause of a wide range of mental disorders.[14] Moreover, the "discovery" of neurotransmitters by neuroscientists played a major role in convincing the public that scientists had found the true cause behind mental illness and other psychological issues.[15]

Whitaker reveals that as an increasing number of new discoveries by scientists fueled the belief in a chemical solution to mental health problems, and more people began to take drugs to cure or manage their psychological problems, the number of people on disability due to mental illness continued to increase: "There were only 50,937 people in state and county mental hospitals with a diagnosis for one of those affective disorders. But during the 1990s, people struggling with depression and bipolar illness began showing up on the SSI and SSDI rolls in ever-increasing numbers, and today there are an estimated 1.4 million people eighteen to sixty-four years old receiving a federal payment because they are disabled by an affective disorder."[16] Whitaker seeks to discover why so many people are now disabled because of mental illness, and what he finds is that sometimes the pills that are supposed to cure people are often making them sicker and less able to function in society.

What is so clear and striking in Whitaker's reporting is that as he unearths study after study showing how and why these drugs end up making people's lives so much worse, he also discovers that virtually every academic, governmental, and medical institution colludes with the promotion of unproven theories and medications:

> But it wasn't just that the AMA [American Medical Association] had given up its watchdog role. The AMA and physicians were also now working with the pharmaceutical industry to promote new drugs. In 1951, the year that the Durham-Humphrey Act was passed, Smith Kline and French and the American Medical Association began jointly producing a television program called The March of Medicine, which, among other things, helped introduce Americans to the "wonder" drugs that were coming to market.[17]

The American Medical Association realized that not only could it help its practicing physicians make more money and increase their prestige by prescribing new psychiatric medications, but the association itself could raise revenue by selling advertisement space to the pharmaceutical companies: "The AMA's revenues from drug advertisements in its journals rose from $2.5 million in 1950 to $10 million in 1960, and not surprisingly, these advertisements painted a rosy picture."[18] Although this funding stream may now look like a small amount of money, the financial relationship between the association regulating the medical profession and Big Pharma has continued to increase, and we can see how science and the medicine has been corrupted by the power of money in the age of neuroliberalism. From a political perspective, the formation of the GUMP complex reveals how the headless, heedless drive of capitalism reshapes all modern liberal institutions as it spreads an amoral mode of human subjectivity.[19]

In case after case, we find that chemists stumbled upon an unexpected effect of a certain chemical, and then pharmaceutical companies figured out how to sell the solution to a mass market; however, rarely did anyone question the social or moral consequences. For example, in the case of Ritalin, the chemical imbalance theory has been used to convince millions of parents to medicate their children, but there is no evidence that anyone knows what really causes attention deficit disorder or even if this diagnosis represents a single disorder or a collection of socially undesirable traits[20]: "Parents were told that children diagnosed with attention deficit hyperactivity disorder suffered from low dopamine levels, but the only reason they were told that was because Ritalin stirred neurons to release extra dopamine."[21] In other words, scientists know that Ritalin causes an increased production of a specific neurotransmitter, but they do not know if this neurotransmitter is related to the disorder. As Whitaker affirms, this method of selling drugs to the public as magical solutions to mental problems has no real scientific basis and represents huge ethical failure on the part of scientists, pharmaceutical companies, universities, and governmental agencies:

> This became the storytelling formula that was relied upon by pharmaceutical companies again and again: Researchers would identify the mechanism of action for a class of drugs, how the drugs either lowered or raised levels of a brain neurotransmitter, and soon the public would be told that people treated with those medications suffered from the opposite problem.[22]

Once again, without the support of scientists, doctors, and neuroliberal ideology, this reductive understanding of the causes and cures of psychological problems would not be possible. It is simply too tempting for medical professionals to resist the power, money, and certainty that comes from being the ones who dispense magical pills, and there has been little accountability or penalty for doctors prescribing medication that ends up being harmful and counter-productive.[23]

To demonstrate how many prescribed drugs do more harm than good, Whitaker looks at several sources of evidence. First of all, he documents that in developing countries with lower access to psychiatric drugs, people recover much faster from depression, bipolar disorder, and other mental problems. He also reveals how in the United States, before the introduction of many of these drugs, most people recovered on their own, and few people were disabled for life. He then turns to studies that document how specific drugs actually destroy parts of the brain and produce an addiction to certain chemicals. What he finds in study after study is that once patients start using a drug, they cannot get off of it because the withdrawal symptoms are so strong that only a return to the drug can make the pain go away. Moreover, several studies have revealed that chemical imbalances are often caused by the medication and not the cause of the mental disturbance:

"Prior to being medicated, a depressed person has no known chemical imbalance. Fluoxetine then gums up the normal removal of serotonin from the synapse, and that triggers a cascade of changes, and several weeks later the serotonergic pathway is operating in a decidedly abnormal manner."[24] The drugs then do affect the brain, but they sometimes do it by transforming the structures in a destructive manner.

In a typical study of the treatment of schizophrenia, we find when comparing patients on psychiatric medications to ones who did not receive medication after hospitalization, the group that was not medicated had a higher recovery rate:

> The off-med group began to improve significantly, and by the end of 4.5 years, 39 percent were "in recovery" and more than 60 percent were working. In contrast, outcomes for the medication group worsened during this thirty-month period. As a group, their global functioning declined slightly, and at the 4.5-year mark, only 6 percent were in recovery and few were working. That stark divergence in outcomes remained for the next ten years.[25]

This study reveals why it is necessary to not rely on short-term research to trace the effects of medication since many of the people who may show immediate positive responses to drugs can end up experiencing long-term harm.[26] However, as we have seen, the GUMP complex motivates psychiatrists and university researchers to buy into the neuroliberal belief that since psychological problems are caused by chemical imbalances inherited through genetic evolution, the only solution is genetic or chemical intervention.

The Psychiatric Narrative

In documenting the link between GUMP and neuroliberalism, Whitaker traces the ways the psychiatric establishment has sought to align itself with a reductive scientific medical model, "the leaders of the APA [American Psychiatric Association] regularly urged the reporters and science writers in attendance to help 'get out the message that [psychiatric] treatment works and is effective, and that our diseases are real diseases just like cardiovascular diseases and cancer,' said APA president Carolyn Robinowitz."[27] In order to center psychiatry on a purely medical model, other forms of therapy, like psychoanalysis, had to be discredited, and this entire process was supported by the pharmaceutical industry.[28] To illustrate this last point, Whitaker describes his visit to an APA conference:

> The best-attended events of the conference were the industry-sponsored symposiums. At every breakfast, lunch, and dinner hour, the doctors could enjoy a sumptuous free meal, which was then followed by talks on the chosen topic. There were symposiums on depression, ADHD, schizophrenia, and the prescribing of antipsychotics to children and adolescents, and nearly all of the speakers hailed from top academic schools. The fact that they all were being paid by the drug companies was openly acknowledged, as the APA, as part of a new disclosure policy, had published a chart listing all the ways that pharmaceutical money flowed to these "thought leaders." In addition to receiving research monies, most of the "experts" served as consultants, on "advisory boards," and as members of "speakers' bureaus."[29]

As a major part of the GUMP complex, the APA helps to solidify the bonds between psychiatrists, universities, and Big Pharma, and much of this bonding has the full support of the government. However, we should not read this situation as a proof of a coordinated conspiracy; rather, the quest for prestige, status, money, and power pushes people to conform to the same amoral capitalistic system. From this perspective, the modern drive to commodify all aspects of life and turn every relationship into an economic exchange is combined with a desire of people to freely conform to a pre-established hierarchy. Therefore, it is not that the alien order of capitalism has simply invaded the pure realm of science, education, medicine, and government: the reality is that the modern scientists, doctors, and public officials have bought into the capitalist drive, and they see no way of escaping from this amoral order.[30]

Whitaker summarizes how the GUMP machine continues to expand and catch more people in its self-destructive web:

> First, by greatly expanding diagnostic boundaries, psychiatry is inviting an ever-greater number of children and adults into the mental illness camp. Second, those so diagnosed are then treated with psychiatric medications that increase the likelihood they will become chronically ill. Many treated with psychotropics end up with new and more severe psychiatric symptoms, physically unwell, and cognitively impaired. That is the tragic story writ large in five decades of scientific literature.[31]

It should be clear now that we have a destructive culture where people of all ages are drugged into disability and shortened lives by esteemed professionals who cash in on the questionable findings of publically funded science.

THE DSM AGAINST PSYCHOANALYSIS

A key factor in adopting the GUMP complex and the neuroliberal drugging of the discontent has been the professional acceptance of the *Diagnostic and Statistical Manual of Mental Disorders* (DSM). As Christopher Lane shows in his book *Shyness*, the authors of this psychiatric bible explicitly sought to eliminate psychoanalytic concepts and theories from mental healthcare as they expanded the number of diagnostic categories to include things that were once considered to be simply aspects of personality.[32] The result is that half of all Americans are now considered to be mentally ill, and for each illness, there is a pharmaceutical solution. Moreover, one reason why the power of the DSM continues to grow is that insurance companies require a diagnosis from the manual in order to ensure payment for treatment and medication. Therefore, psychiatrists, psychoanalysts, and other mental health practitioners have to buy into the DSM even if they do not believe in its categories and underlying theories.

Lane points out that the DSM has been globalized and other system of diagnosis more sympathetic to psychoanalysis have been discredited. In tracing the development of the recent additions of the manual, he depicts how in the quest to ground mental disorders in a biomedical model, the writers of the DSM had to return to categories developed before Freud, and this is one of the ways they have acted to repress psychoanalysis.[33] Lane explains that when the DSM-III was being formulated, the lead author, Robert Spitzer, purposely selected experts who were against psychoanalysis. One reason for this decision was that due to the rising costs of mental healthcare in the 1980s, the APA felt that a more scientific foundation for diagnosis and treatment was necessary. Since the general public tends to believe in medicine and the empirical sciences, the biomedical model had to be privileged over the more abstract and difficult-to-understand theories of psychoanalysis. In fact, Spitzer worked hard to erase the word neurosis from the manual, and instead he used the term disorder because it sounded more medical and scientific.

It is important to stress that this all-powerful book contains the words "statistical" and "disorders" in its title because one of the pain goals was to cash in on the prestige of respected science. As Lane points out, the authors of the DSM-III were greatly influenced by researchers from the Washington University psychiatry department who argued that mental disorders should be defined in ways that are "descriptive, explicit, and rule-driven."[34] In other words, the push for scientific-sounding diagnostic categories worked against the more abstract and non-empirical aspects of the unconscious, psychoanalysis, and the drives. Furthermore, Lane argues that the diagnosis of mental health issues usually involves a focus on manifest symptoms, but from a psychoanalytic perspective, symptoms are always the result of repression and distortion. It is also clear that the DSM system motivates practitioners to treat symptoms and not the underlying causes.

Lane argues that psychological symptoms may indicate social deviance or discontent, but the DSM removes the role of society and the family from its focus. For instance, by medicalizing shyness, mental health specialists end up pathologizing people who are simply unhappy or different. Not only then is discontent being drugged and silenced, but our social system and individual responsibility are removed from concern since all problems end up having a biomedical foundation. A central driving force behind GUMP is thus derived from several interlocking forces: insurers need standardized diagnostic categories in order to control healthcare costs and regulate treatment plans, and in order to render diagnosis more explicit and rule-based, the designers of the manual have to repress psychoanalysis and focus on symptoms that follow a biomedical model. In turn, this need for a biomedical model sends the writers of the manual to university researchers who are often funded by the federal government and the pharmaceutical corporations. The GUMP complex therefore employs neuroliberal ideology to justify the drugging of discount and the marketing of mental health as it also works to repress psychoanalysis and the unconscious.

A New Anti-psychiatry?

As attempts at countering this growing dominance of GUMP and neuroliberalism, Darian Leader's *Strictly Bipolar* and Paul Verhaeghe's *What about Me?* return to a psychoanalytic understanding of psychological problems.[35] Leader opens his work by stressing that the bipolar diagnosis has recently increased in America from 1% of the population to 25%, while psychiatric prescriptions for children has increased 400%.[36] One reason for this incredible increase in diagnosis and medication can be traced to the fact that in the mid-1990s, patents on mainstream anti-depressants had expired, and pharmaceutical corporations needed a new pathology and set of drugs to revamp their business models.[37] As part of this process, these companies funded Internet sites and journal articles dedicated to describing and diagnosing bipolar disorders, and one of the results of this process was that many people who had been diagnosed as being depressed were rediagnosed as being bipolar.[38] By changing the diagnosis, psychiatrists, scientists, and Big Pharma were able to argue that the reason why the older forms of medication often did not work was that they were wrongly prescribed, but now with a proper diagnosis of bipolar disorder, new medications were available.[39]

Of course a major issue exposed here is that once mental health becomes a lucrative business, decisions are often made based on profit and not care.[40] Furthermore, in helping to develop new forms and levels of bipolarity, university scientists, psychiatrists, and governmental agencies were able to increase their influence and funding as they provided a false solution to a new epidemic. Here, we see how GUMP not only describes how drugs are marketing to a vulnerable public but also how universities, doctors, and public institutions have been corrupted by the combination of science and capitalism. Since professors need grants to perform their research, and these grants often are supported by vested interests, the scientific search for objective truth is tainted by the need for money.[41] Thus, the modern division between pure science and capitalist self-interest is lost as universities, corporations, and governments collude in order to sell the new solution to a new problem. For example, as Leader points out, at the very moment the patents were running out on anti-depressants, Depakote received a patent to treat mania.[42] Likewise, since Lithium could not be patented, because it was a naturally occurring element, new drugs had to be developed to take its place.[43]

Due in part to the continual increase in diagnostic categories in the DSM and the availability of new medications to treat every disorder, many psychiatrists and other mental health practitioners now see their jobs transformed from being therapists to being medication managers.[44] Once again, this turn to a chemical solution is a win-win-win proposition for doctors because it allows them to see more patients and receive more money while shoring up their roles as the ones who know the truth about mental disorders. Moreover, as all of these drugs produce side effects, patients are increasingly given multiple medications to deal with their disorders, and this further increases the role and the money of prescribing physicians. Leader points out that while psychoanalysis seeks to deal with thoughts, feelings, and behaviors, the new pharmacological model works by managing symptoms. For the practicing psychiatrist, this transition from analysis to medication can make the job much more efficient and confortable as one no longer has to listen to patients to understand their problems; instead, the psychiatrist simply has to manage medication regimes.

MEDICATING NEOLIBERALISM

In Verhaeghe's book, this medicalization of mental disorders is placed within the context of neoliberal politics and culture. He argues that a new form of social Darwinism has emerged, which posits that we are controlled by our selfish genes and neurons, and that in a meritocratic culture, the people who work the hardest and have the most talent will be justly rewarded. Through this naturalization of the social hierarchy, people only have themselves and their genes to blame for their lack of resources and recognition, and if they are unhappy, there is always a drug to take to enhance their performance or repress their anxiety.

On the one hand, for Verhaeghe, neoliberalism endorses a biological model for human development, and on the other hand, it continues to blame individuals for their failure to succeed in a fair and open society.[45] This conflict between two uses of biology results in the ideological contradiction of arguing at the same time that free market capitalism is the natural order of society, and people are preprogrammed by natural selection to succeed or fail in society. In this structure, the individual becomes the locus of responsibility, while the individual is also seen as determined by uncontrollable biological forces. The solution then to this contradiction is to turn to medication and away from social intervention.

Verhaeghe argues that as attacks on the welfare state are enhanced, healthcare becomes focused on measurable outcomes, and this results in a constant effort to measure, rank, rate, and evaluate all aspects of medical care. In this context, the proliferation of diagnostic categories and chemical solutions makes perfect sense since the stress is on what can be measured and not on what really helps people who are suffering. Furthermore, with this combination of science and capitalism, medical professionals are incentivized to see people on purely biological terms.[46] At the same time, due to the public disinvestment in higher education, universities are also being constantly rated, ranked, and measured, and this pushes them to increase their research output and rely more on external funding. As capitalism begins to increase its hold on both hospitals and universities, these institutions are transformed into regulated businesses that must serve economic interests and are structured by an anti-social culture of constant competition and anxiety.

Verhaeghe argues that mental disorders are always disorders in relation to specific social norms, and therefore, even if there is a biological foundation to a pathology, that pathology has to be understood within a cultural context.[47] However, it should be clear that the neuroliberal medical model of mental disturbances and normal mental functioning works by repressing and foreclosing the role played by language, society, culture, and subjectivity, and in this sense, this type of scientism results in repressing the unconscious and psychoanalysis itself.

Verhaeghe affirms that the push to measure everything in science and medicine has resulted in a situation where alternative therapies, like psychoanalysis, can no longer find a place at universities or hospitals.[48] Meanwhile, the DSM increased the number of diagnostic categories from 180 in the second edition to 365 in the fourth edition, and as the number of entries expanded, the focus on a neurobiological causality was also enhanced.[49] What most surprises Verhaeghe in this neuroliberal context is that this explosion in medicating discontent has not been coupled with any strong protests from within or outside of the psychiatric community.[50] After all, in the 1970s, we witnessed many public outcries against over-medicalization and the social construction of mental illness, but during the current expansion of this model, only a few groups outside of psychiatry have critiqued the system.[51] Furthermore, Verhaeghe points out that a low number of patients resist this new model, and this may be because they buy into the scientific rhetoric and advertising that promotes drugs as the solution to mental diseases.[52] He adds that in the age of the Internet, people often go online and show up to the doctor's office with their own self-diagnosis, and this gives them a sense of identity and control.[53] From Verhaeghe's perspective, as we move into a post-tradition society of unstable identifications, a diagnostic label might seem like the best way for people to stabilize their identities.[54]

Making matters worse is the fact that children are being diagnosed and medicated at much higher rates and at much earlier ages. For instance, teachers are now referring "difficult" students to medical practitioners, and parents, who worry that their children are not succeeding in school, are seeking medical solutions to behavioral problems.[55] It therefore appears that all levels of society have bought into the disease model, which ignores cultural and economic forces, and relates all mental disturbances to a neurobiological explanation. Yet, study after study shows that there is very little proof for the scientific theories that support both the new diagnostic categories and the related drug treatments.[56]

For Verhaeghe, the basic problem is that a dominant scientific ideology has taken hold with very little resistance, and no one wants to talk about the possible social causes for anxiety, depression, or confusion. Thus, as neoliberal society undermines social trust and increases economic inequality and employment uncertainty, the focus is placed on the failing individual who is shaped by genes and neurons.

Freud Against GUMP

From a psychoanalytic perspective, these biochemical models fail to take into account the Freudian theories of the unconscious, symptoms, sexuality, fantasies, transference, and drives. Moreover, Freud's theories and concepts were driven by the development of his technique, and it is on this level that we see the radical conflict between psychoanalysis and GUMP. Although Freud began as a neurologist and soon turned to hypnosis in order to suggest the missing memories to his hysterical patients, throughout the development of his technique, he moved away from the power of the physician as he focused on letting the patient free associate with little intervention from the doctor.[57] After his original hypnotic method failed, Freud realized that symptoms would only go away for good if the patients came up with the missing information on their own.[58] By clinging to the neutrality of the analyst and the free associations of the patient, Freud problematized the idealization of the doctor that often dominates in the realm of medicine and science today. In fact, Freud posited that real science is based on the scientist recognizing the limitations of knowledge: "Only a man who really knows is modest, for he knows how insufficient his knowledge is."[59] Freud used this ethical perspective to argue against the physicians and the scientists who approach each subject from a perspective of being the one who knows. Furthermore, Lacan later employed this same ethical idea to posit that transference involves falling in love with the person we suppose knows the truth.[60] This psychoanalytic perspective thus challenges the view of many of the scientists and medical practitioners who dominate the fields of mental health today because the respect for science and medicine tends to unleash a wild form of transference that can be seen in the power of placebos and the ability of the medical and pharmaceutical establishment to convince millions of people that they have the solution to all mental problems.[61]

Freud was able to see why so many psychoanalysts buy into the medical model because he had to deal with many of his followers who argued that people without medical degrees should not be allowed to practice analysis:

> How can I explain the attitude of these pupils of mine to you? I do not know for certain; I think it must be the power of professional feeling. The course of their development has been different from mine, they still feel uncomfortable in their isolation from their colleagues, they would like to be accepted by the 'profession' as having plenary rights, and are prepared, in exchange for that tolerance, to make a sacrifice at a point whose vital importance is not obvious to them. Perhaps it may be otherwise; to impute motives of competition to them would be not only to accuse them of base sentiments but also to attribute a strange shortsightedness to them ... These pupils of mine may be influenced by certain factors which guarantee a doctor an undoubted advantage over a layman in analytic practice.[62]

Freud anticipated here what Whitaker describes, which is that in a quest for certainty, prestige, money, and influence, psychiatrists and other mental health professionals have turned to a destructive chemical solution to all psychological disturbances. As Freud stated above, the desire for professional status also served to repress psychoanalysis in favor of medicine because of its higher social status and special rights. At the same time, the biomedical model for mental disturbances helps to position the patient as a helpless victim of inherited chemicals and genes, and this feeds into fantasies concerning the innocent individual who can do no harm and cannot be criticized.[63]

In his text, "The Question of Lay Analysis," Freud offered a sustained criticism of the medicalization of psychological disorders, and his critique can be seen as offering an effective counter-discourse to the use of neuroliberalism in the GUMP complex:

> The first consideration is that in his medical school a doctor receives a training which is more or less the opposite of what he would need as a preparation

for psycho-analysis. His attention has been directed to objectively ascertain-able facts of anatomy, physics and chemistry, on the correct appreciation and suitable influencing of which the success of medical treatment depends. The problem of life is brought into his field of vision so far as it has hitherto been explained to us by the play of forces which can also be observed in inanimate nature. His interest is not aroused in the mental side of vital phenomena; medicine is not concerned with the study of the higher intellectual functions, which lies in the sphere of another faculty. Only psychiatry is supposed to deal with the disturbances of mental functions; but we know in what manner and with what aims it does so. It looks for the somatic determinants of men-tal disorders and treats them like other causes of illness.[64]

Freud's critique of the medical focus on the somatic causes for mental ill-ness is still relevant today since one of the main reasons why psychiatrists and other physicians have bought into the pharmaceutical solution is that they follow neuroscience and evolutionary psychology in the pursuit of tying all mental processes to inherited biological programs.[65]

As Freud argued, psychiatry is often seduced by the simple idea that all mental processes have a biological foundation, and all disturbances can be examined by using the same scientific methodologies that are applied to inanimate objects. Although Freud was consistent in his refusal to see medical education as the main part of the training for psychoanalysts, he held out hope that some day science would provide empirical evidence to support his research:

> Neurotics, indeed, are an undesired complication, an embarrassment as much to therapeutics as to jurisprudence and to military service. But they exist and are a particular concern of medicine. Medical education, however, does nothing, literally nothing, towards their understanding and treatment. In view of the intimate connection between the things that we distinguish as physical and mental, we may look forward to a day when paths of knowledge and, let us hope, of influence will be opened up, leading from organic biol-ogy and chemistry to the field of neurotic phenomena. That day still seems a distant one, and for the present these illnesses are inaccessible to us from the direction of medicine.[66]

In response to this desire for a chemical and biological understanding of neurosis, some may argue that neuroscience and other related fields have now provided the empirical evidence for which Freud was looking; however, as we have seen the neuroliberal understanding of the unconscious has served to further repress psychoanalysis and the unconscious.[67]

One reason why Freud did not want psychoanalysts to be trained at medical schools was that he accurately predicted the repression of the unconscious:

> It would be tolerable if medical education merely failed to give doctors any orientation in the field of the neuroses. But it does more: it gives them a false and detrimental attitude. Doctors whose interest has not been aroused in the psychical factors of life are all too ready to form a low estimate of them and to ridicule them as unscientific. For that reason they are unable to take anything really seriously which has to do with them and do not recognize the obligations which derive from them. They therefore fall into the layman's lack of respect for psychological research and make their own task easy for themselves.[68]

What Freud feared has come to fruition: medically trained psychoanalysts and psychiatrists have taken the easy way out by turning to medication and reductive scientific explanations in their efforts to deal with neurosis and other psychic pathologies.

As I have argued throughout this book, we must always approach science as being embedded in social and political ideologies, and in the age of neoliberalism, this attention to the social construction of scientific discourses becomes even more vital. The power of the new brain sciences to shape public and professional opinions has resulted in a situation where people are told that they have only themselves to blame for their social and psychological problems, and yet, they are also told that most of their mental processes have been preprogrammed by evolution. neuroliberalism then sends a contradictory message of individual freedom and biological control, and one of the main ways people deal with this conflict is by turning to medication to reprogram their inherited genetic instincts. In this context, psychoanalysis represents an essential counter-discourse.

NOTES

1. Karl Marx and Friedrich Engels, *Selected Correspondences 1846–1895* (London: Lawrence and Wishart, 1943): 125–6.
2. Dawkins, Richard. *The selfish gene.* Oxford university press, 2016.

3. Frances, Allen. "Saving normal: An insider's revolt against out-of-control psychiatric diagnosis, DSM-5, big pharma and the medicalization of ordinary life." *Psychotherapy in Australia* 19.3 (2013): 14.

4. Pollock, Anne. "Transforming the critique of Big Pharma." *BioSocieties* 6.1 (2011): 106–118.

5. Sharfstein, Steven S. "Big pharma and American psychiatry." *The Journal of nervous and mental disease* 196.4 (2008): 265–266.

6. Gøtzsche, Peter C., Richard Smith, and Drummond Rennie. *Deadly medicines and organised crime: how Big Pharma has corrupted healthcare.* London, UK: Radcliffe, 2013.

7. Washburn, Jennifer. *University, Inc.: The corporate corruption of higher education.* Basic Books, 2008.

8. Mulvany, Julie. "Disability, impairment or illness? The relevance of the social model of disability to the study of mental disorder." *Sociology of Health & Illness* 22.5 (2000): 582–601.

9. Freud, Sigmund. *On narcissism: An introduction.* Read Books Ltd, 2014.

10. Whitaker, R. (2011). *Anatomy of an epidemic: Magic bullets, psychiatric drugs, and the astonishing rise of mental illness in America.* Broadway.

11. Ibid., xi.

12. Ibid.

13. Ibid., 4.

14. Lexchin, Joel, et al. "Pharmaceutical industry sponsorship and research outcome and quality: systematic review." *Bmj* 326.7400 (2003): 1167–1170.

15. Moncrieff, J. (2008). *The myth of the chemical cure: A critique of psychiatric drug treatment.* Macmillan.

16. Whitaker, 7.

17. Ibid., 56.

18. Ibid., 57.

19. Hedges, Chris. *Death of the liberal class.* Vintage Books Canada, 2011.

20. Hinshaw, Stephen P., and R. Scheffler. "The ADHD explosion." *Myths, Medication, and Today's Push for Performance. New York: Oxford UP* (2014).

21. Whitaker, 77.

22. Ibid.

23. Miller, Rosalind S., George H. Wiedeman, and Louis Linn. "Prescribing Psychotropic Drugs: Whose Responsibility?." *Social work in health care* 6.1 (1981): 51–61.

24. Ibid., 81.

25. Ibid., 115.

26. One of the strongest indications that our current pharmacological model is not working is in the treatment of people diagnosed with schizophrenia:

"In 1955, there were 267,000 people with schizophrenia in state and county mental hospitals, or one in every 617 Americans. Today, there are an estimated 2.4 million people receiving SSI or SSDI because they are ill with schizophrenia" (93). Once again, no one wins in this system that places millions of people on public assistance in part because the prescribed medications are disabling them.

27. Whitaker, 172.
28. Whitaker, R., & Cosgrove, L. (2015). *Psychiatry under the influence: Institutional corruption, social injury, and prescriptions for reform*. Palgrave Macmillan.
29. Whitaker, 173.
30. Suciu, Titus. "IS CAPITALISM ETHICAL?" *Bulletin of the Transilvania University of Brasov. Economic Sciences. Series V 2* (2009): 237.
31. Whitaker, 209.
32. Lane, C. (2008). *Shyness: How normal behavior became a sickness*. Yale University Press.
33. American Psychiatric Association. (1980). *DSM-III-R: Diagnostic and statistical manual of mental disorders*. American Psychiatric Association.
34. Lane, 68.
35. Leader, D. (2013). *Strictly bipolar*. Penguin UK. Verhaeghe, P. (2014). *What about Me?: the struggle for identity in a market-based society*. Scribe Publications.
36. Leader, 1.
37. Ibid., 4.
38. Ibid.
39. Ibid.
40. Frances, A. (2013). Saving normal: An insider's revolt against out-of-control psychiatric diagnosis, DSM-5, big pharma and the medicalization of ordinary life. *Psychotherapy in Australia, 19*(3), 14.
41. Washburn, J. (2008). *University, Inc.: The corporate corruption of higher education*. Basic Books.
42. Leader, 5.
43. Ibid., 5.
44. Ibid., 10.
45. McNamee, S. J., & Miller, R. K. (2009). *The meritocracy myth*. Rowman & Littlefield.
46. Verhaeghe, 130–32.
47. Ibid., 105.
48. Ibid., 139.
49. Ibid., 182.
50. Ibid.
51. Ibid.

52. Ibid., 183.
53. Ibid.
54. Giddens, A. (1993). Life in a Post-Traditional Society. *REVISTA DE OCCIDENTE*, (150), 61.
55. Mayes, R., Bagwell, C., & Erkulwater, J. L. (2009). *Medicating children: ADHD and pediatric mental health.* Harvard University Press.
56. Verhaeghe, 191.
57. Freud, Sigmund. "The dynamics of transference." *Classics in Psychoanalytic Techniques* (1912).
58. Freud, S. (1914). Remembering, repeating and working-through (Further recommendations on the technique of psycho-analysis II). *Standard edition, 12*, 145–156.
59. Freud, S. (1959). *The Question of Lay Analysis.* In *The Standard Edition of the Complete Psychological Works of Sigmund Freud, Volume XX (1925–1926): An Autobiographical Study, Inhibitions, Symptoms and Anxiety, The Question of Lay Analysis and Other Works* (231).
60. Lacan, J. (1998). *The four fundamental concepts of psycho-analysis* (Vol. 11). WW Norton & Company: 123–5.
61. Shapiro, A. K. (1971). Placebo effects in medicine, psychotherapy, and psychoanalysis. *Handbook of psychotherapy and behavior change. New York: Wiley,* 21.
62. Freud, *The Question*, 138–9.
63. Cole, Alyson Manda. *The cult of true victimhood: from the war on welfare to the war on terror.* Stanford University Press, 2007.
64. Ibid., 229.
65. Moncrieff, Joanna. *The myth of the chemical cure: A critique of psychiatric drug treatment.* Macmillan, 2008.
66. Freud, *The Question*, 230.
67. Levin, F. M. (2003). *Mapping the mind: The intersection of psychoanalysis and neuroscience.* Karnac Books.
68. Freud, S., *The Question*, 230.

CHAPTER 7

Conclusion

Abstract The main goal of this conclusion is to show how the new brain sciences are so affected by neoliberal ideology that we can consider many of them to be political theories masquerading as scientific discoveries. Just as neoliberalism stresses the role of the private individual adapting to an unequal social system, neuroliberalism focuses on how evolution has pre-programmed us to pursue our self-interest by trying to outcompete others. In this context, capitalist markets are seen as natural structures dedicated to sorting out winners and losers in a cultural form of evolutionary selection. Likewise, meritocracy can be viewed as the liberal version of this naturalized competitive order where people are rewarded for their learned or natural talents instead of their inherited wealth.

Keywords Meritocracy • Neoliberalism • Ideology • Brain sciences • Evolution

The main goal of this book has been to show how the new brain sciences are so affected by neoliberal ideology that we can consider many of them to be political theories masquerading as scientific discoveries. Just as neoliberalism stresses the role of the private individual adapting to an unequal social system, neuroliberalism focuses on how evolution has preprogrammed us to pursue our self-interest by trying to outcompete others. In this context, capitalist markets are seen as natural structures dedicated to sorting out winners and losers in a cultural form of evolutionary selection.

© The Author(s) 2017
R. Samuels, *Psychoanalyzing the Politics of the New Brain Sciences*,
https://doi.org/10.1007/978-3-319-71891-0_7

Likewise, meritocracy can be viewed as the liberal version of this natural-ized competitive order where people are rewarded for their learned or natural talents instead of their inherited wealth.

In my reading of Steven Pinker's *Blank Slate* and other works by evolu-tionary psychologists, I pointed out that much of their publications deal with discrediting the social sciences, psychoanalysis, cultural theory, pro-gressive parenting, and welfare state programs. Not only do these neuro-liberal discourses want to show how their approach to understanding human nature is superior to the humanities and traditional social sciences, but they also desire to discredit progressive attempts to intervene in soci-ety to make the world more fair and just. From the perspective of these new brain sciences, since we are determined by our genes to act in certain programmed ways, there is no ability to go against universal human nature.

To counter these neuroliberal discourses, I have turned to psychoanaly-sis to show how the Freudian conceptions of sexuality and the uncon-scious reveal the limitations of neuroscience, evolutionary psychology, and behavioral economics. As Freud discovered throughout his work, humans often engage in self-destructive behavior that counters the evolutionary focus on self-regulation and survival. Moreover, the ability of individuals to combine pleasure with pain and fiction with reality opens a space for disrupting biological determinism. Starting with his first work with hys-terical patients, Freud discovered that not only did their physical symp-toms not make anatomical sense, but the patients themselves did not know the origin or causes for their mental illnesses.

As we saw in the last chapter, Freud anticipated what would happen if psychoanalysis was absorbed into the medical world. By repressing the unconscious and committing to a biological explanation for mental pro-cesses, psychiatrists and psychoanalysts have bought into the Government University Medical Pharmaceutical complex. One of the results of this system is that social discontent is medicalized, and drugs are seen as the main option for mental discomfort. Moreover, as neoliberal society pushes people to compete for scarce resources in an unequal and unfair society, which blames individuals for their failures, people turn to medication to relieve their stress and anxiety. In fact, citizens are told that if they want to succeed in the competitive rat race, they should take performance enhance-ment drugs that will help them focus and work long hours.

The irony is that as neoliberal culture undermines the welfare state and other public institutions, people are informed that they only have them-selves to blame, and yet, they are also told that it is their genes and neurotransmitters that are actually running the show. The ideology of individualism is thus coupled with an anti-individual discourse of biological determinism. Like the premodern belief in predestination, neuroliberalism tells us that we will be rewarded for our individual efforts and talents, but all of our mental processes have been preprogrammed through natural selection. The paradox of this cultural moment, then, is that the celebration of the individual over society is coupled with a belief in the inability of the individual to overcome universal human nature.

Instead of seeing the new brain sciences as being objective, neutral, and unbiased, I have stressed that their vision of human nature is deeply political and historical, and one reason why neuroliberal discourses may be so bent on undermining the social sciences and the humanities is that they want to eliminate any discourse that can call them into question. As Lacan always insisted, in any discourse, we must ask who is really speaking and who are they really speaking to? Although scientists want to believe that they are speaking to everyone from a position of no one, we have seen that the new brain sciences are full of assumptions and partial views that just happen to reinforce the social and economic status quo.

INDEX[1]

B

Behavioral economics, 1, 4, 6, 15, 19,
22, 27, 31, 38, 59–80, 86, 94,
99, 104, 110, 115, 117, 138

C

Capitalism, 12, 62, 69–71, 78, 116,
121, 124, 127–129, 137
Conformity, 10–12, 16, 17, 24, 26,
60, 70–80, 82n48
Cynical conformity, 24, 26, 72, 76,
77

D

Damasio, Antonio, 1, 2, 6n2, 9–31,
35, 60, 65, 68, 94
Death drive, 11
Descartes, Rene, 6n2, 9, 21–26,
28–30, 31n1, 44, 48

E

Evolutionary psychology, 1, 4, 6, 6n3,
15, 19, 21, 22, 27, 35–55, 60,
68, 71, 80, 86, 88, 90, 92–97,
102, 110, 111n19, 111n20,
112n34, 112n35, 115, 132, 138

F

Free market, 4, 17, 43, 46, 53, 60, 75,
77, 116, 128
Freud, Sigmund, 112n31, 134n9
Beyond the Pleasure Principle, 11
Civilization and Its Discontents, 14
"Dynamics of the Transference",
130
"The Economic Problem of
Masochism", 94
"The Ego and the Id", 15, 108
"Formulations on the two principles
of mental functioning", 10

[1]Note: Page numbers followed by 'n' refers to notes.

© The Author(s) 2017
R. Samuels, *Psychoanalyzing the Politics of the New Brain Sciences*,
https://doi.org/10.1007/978-3-319-71891-0